Between the years 1890 and 1924, the dominant view of the Universe suggested a cosmology largely foreign to contemporary ideas. First, astronomers believed they had confirmed that the Sun was roughly in the center of our star system, the Milky Way Galaxy. Second, considerable evidence indicated that the size of the Galaxy was only about one-third the value now accepted by today's astronomers. Third, it was thought that interstellar space was completely transparent, that there was no absorbing material between the stars. Fourth, astronomers believed that the Universe was composed of numerous star systems comparable to the Milky Way Galaxy. The method that provided this picture and came to dominate cosmology was "statistical" in nature, because it was based on the counts of stars and their positions, motions, brightnesses, and stellar spectra.

Drawing on previously neglected archival material, Professor Paul describes the rise of this statistical cosmology in light of developments in nineteenth-century astronomy and explains how this cosmology set the stage for many of the most significant developments we associate with the astronomy of the twentieth century. Statistical astronomy was the crucial link that provided much of modern astronomical science with its foundation.

The Milky Way Galaxy and
Statistical Cosmology,
1890–1924

The Milky Way Galaxy and Statistical Cosmology, 1890–1924

ERICH ROBERT PAUL
Dickinson College

CAMBRIDGE
UNIVERSITY PRESS

Published by the Press Syndicate of the University of Cambridge
The Pitt Building, Trumpington Street, Cambridge CB2 1RP
40 West 20th Street, New York, NY 10011-4211, USA
10 Stamford Road, Oakleigh, Melbourne 3166, Australia

First published 1993

Printed in the United States of America

Library of Congress Cataloging-in-Publication Data
Paul, Erich Robert, 1943–
The Milky Way galaxy and statistical cosmology, 1890–1924 / Erich
Robert Paul.
p.   cm.
Includes bibliographical references and index.
ISBN 0 521 35363 7
1. Milky Way – History.   2. Cosmology – Statistical methods –
History.   3. Astronomy – Statistical methods – History.   I. Title.
QB857.7.P38   1993
523.1'13 – dc20   92–36530
                                                      CIP

A catalog record for this book is available from the British Library.

ISBN 0 521 35363 7 hardback

To KATHLEEN and our children,
ANN-MARIE, LISA, JULIET, ERICA,
and CHRISTOPHER – for all your
love and support

CONTENTS

ILLUSTRATIONS

ACKNOWLEDGMENTS

Numerous colleagues have contributed to this book by giving generously their advice and criticism. With deepest gratitude I acknowledge my indebtedness to the late Victor E. Thoren of Indiana University, whose assistance was invaluable as I came to understand the history of astronomy and later commenced to research and write this study; his interest in the evolution of this work has been a source of continuing stimulation. Robert W. Smith of The Johns Hopkins University and the Smithsonian Institution has been particularly generous in his reading of the entire manuscript and in making suggestions that have given me considerable pause. I am also grateful to Michael J. Crowe of the University of Notre Dame and to Steven J. Dick of the U.S. Naval Observatory for reading various chapters and providing detailed comments. Michael A. Hoskin of the University of Cambridge and Owen Gingerich of Harvard University, in their capacity as editor and associate editor of the *Journal for the History of Astronomy*, provided valuable editorial assistance on earlier versions of Chapters 4, 6, and 8. I thank Felix Schmeidler of the Munich Observatory and Adriaan Blaauw of the Kapteyn Astronomical Laboratory, both senior research astronomers, and historian Ute Lindgren at the Institute for the History of Science at the University of Munich, for their assistance with the location of esoteric sources.

I am additionally grateful to other professional colleagues and friends, both historians and astronomers, particularly Ronald Brashear, David DeVorkin, C. Robert Ferguson, Norris Hetherington, John Lankford, Donald Osterbrock, and Lewis Pyenson, who provided assistance with various aspects of this study.

Without the generous assistance of many archivists and librarians, it would have been impossible to examine the manuscript sources on which this study is based. Thanks are due to the

Huntington Library and the Mount Wilson and Las Campanas Observatories (George Ellery Hale Papers), Kapteyn Astronomical Laboratory and the University of Groningen (J. C. Kapteyn Papers), Universitätsbibliothek Heidelberg (Max Wolf Papers), Sternwarte München (Hugo von Seeliger Papers), Widener Library, Harvard University Archives (Harlow Shapley Papers), American Philosophical Society (Karl Schwarzschild Microfilm Collection), Center for History of Physics at the American Institute of Physics (Bart Bok Interview), Yerkes Observatory Archives (Kapteyn–Parkhurst Papers and Walter S. Adams Papers), Library of Congress (Henry Smith Pritchett and Simon Newcomb Papers), Stockholm University Library (Hugo Gyldén Papers), and Lowell Observatory Archives ("Early Correspondence of the Lowell Observatory").

I am also particularly grateful to the following libraries and archives: Indiana University Library, Bloomington; Library of the U.S. Naval Observatory, Washington, D.C.; University of Pennsylvania Library; Boyd L. Spahr Library of Dickinson College; Bayerischen Akademie der Wissenschaften; Bayerisches Hauptstaatsarchiv München; Bayerische Staatsbibliothek München; Universitätsbibliothek München; Archiv Universität München; Deutsches Museum, Munich; Stadtbibliothek München; Archiv der Münchener Sternwarte; and the Niedersachsische Staats- und Universitätsbibliothek Göttingen.

Permission has been received to quote from the following archival sources: Mount Wilson and Las Campanas Observatories (George Ellery Hale Papers), Kapteyn Astronomical Laboratory and University of Groningen (J. C. Kapteyn and David Gill Papers), Harvard University Archives (Harlow Shapley Papers), and the Lowell Observatory Archives. I am grateful to Michael Hoskin, editor of the *Journal for the History of Astronomy*, for permission to reproduce material from my previous articles. I am indebted to Helen Wheeler, my editor with Cambridge University Press, who has provided invaluable aid in preparation of the final manuscript.

My research has been partially supported by the U.S. National Science Foundation under grants SES 82-06416 and SES 85-09508, by a National Endowment for the Humanities grant, and by a Deutsches Museum summer grant.

Finally, I am grateful to Dickinson College for two Board of Advisers research grants, the second of which provided additional financial means necessary to complete this book during a recent year-long sabbatical in the Wasatch Mountains. Kathleen and our children, Ann-Marie, Lisa, Juliet, Erica, and Christopher, supported me during many hours of solitary reflection, occasionally providing relief as we explored issues of cosmic significance at Park City, Snowbird, and Alta.

Although I am indebted to the wisdom of all these friends and colleagues, responsibility for any errors of fact or judgment that occur is entirely mine.

ABBREVIATIONS OF MANUSCRIPT SOURCES

In the footnotes to chapters, manuscript sources are abbreviated as follows:

A.I.P.	Center for History of Physics, American Institute of Physics, New York City
B.H.A.	Bayerisches Hauptstaatsarchiv, Munich
D.M.	Deutsches Museum, Munich
Göttingen	Universitätsbibliothek Göttingen, Germany
Hale	George Ellery Hale Papers, Huntington Library, San Marino, California
Heidelberg	Universitätsbibliothek Heidelberg, Germany
K.A.L.	Kapteyn Astronomical Laboratory, Groningen, Holland
Lowell	Lowell Observatory Archives, Flagstaff, Arizona
S.B.M.	Stadtbibliothek, Munich
Schwarzschild	Schwarzschild Microfilm Collection, American Philosophical Society, Philadelphia
Shapley	Widener Library, Harvard University
Stockholm	Stockholm University Library
Yerkes	Yerkes Observatory Archives, Williams Bay, Wisconsin

INTRODUCTION

This book concerns itself with pre-relativistic cosmology and with those areas of astronomy that contributed to our understanding of the structure of the Universe as it developed before about 1920. Since the time of Isaac Newton in the seventeenth century, two general trends seem to have predominated among those who speculated upon and investigated galactic astronomy as cosmology. Some astronomers, such as Thomas Wright, Immanuel Kant, and Johann Lambert in the eighteenth century, and Maxwell Hall, Cornelis Easton, and Eduard Schönfeld in the nineteenth century, developed stellar and galactic models of the cosmos largely without concern for such microproblems as stellar position, stellar movement, and the brightness of stars. Others believed that a full-blown cosmology first requires close attention to the observational details of the stellar universe.

Although these approaches are complementary, they are also radically different and produce quite different results. On the one hand, cosmology in the first sense has often been more speculative than substantive and has relied heavily on philosophical assumptions as a substitute for greater empirical detail. Thus there was a far greater concern with conceptual problems dealing with, for example, cosmogony, Newtonian universal gravitation, and even the plurality of worlds, than with empirical and observational problems. In the more restricted sense, cosmology has been less global, but, in the long run, perhaps more productive astronomically. As the nineteenth-century Italian astronomer Giovanni Celoria put it succinctly: "To speculate upon the structure of the Milky Way without a complete knowledge of its features is to build upon the void."[1] "Statistical astronomy," the technical

---

[1] Giovanni Celoria, quoted in M. Turchetta and G. Gavazzi, "Nineteenth-Century Italian Contributions to Galactic Theory," *Journal for the History of Astronomy*, 18(3) (August 1987), 207 (196–208).

subdiscipline that constitutes the present discussion, is quint-
essentially representative of the latter approach.

Historically, statistical astronomy had been of crucial importance to developments in pre-relativistic cosmology. For during
the first several decades of the present century astronomers generally believed that a statistical approach to analyzing the aggregate
of stars would eventually lead to an accurate understanding of
the architecture of the stellar universe. Although William Herschel
initiated this program late in the eighteenth century, it was not
until a century later that it began to achieve significant promise
with the work of two prominent astronomers, the Dutchman
J. C. Kapteyn (1851–1922) and the German Hugo von Seeliger
(1849–1924). Roughly from 1890 until the early 1920s, Kapteyn
and Seeliger shaped astronomers' cosmological views of the stellar
universe.

The basic contributions that Seeliger and Kapteyn made to this
field of astronomy during the three decades prior to the "astronomical revolution" of the 1920s was enormous. Theirs was not
a singular project; many contributed to this research tradition
focusing on empirical, conceptual, and methodological problems.
Not only did both Kapteyn and Seeliger develop centers for astronomical research, but programs contributing to this particular
problem in statistical astronomy also emerged under the direction
of C. V. L. Charlier in Sweden and Edward Pickering at Harvard.
Important advances were also contributed by Karl Schwarzschild
in Germany, Arthur S. Eddington and Frank W. Dyson in England, and Heber D. Curtis and George Ellery Hale in the United
States. Principally under the direction of these astronomical leaders,
lesser known astronomers made their careers by contributing much
also. In the years immediately following 1900 these groups developed well-defined centers of research: Critical problems were
clearly identified; methodological approaches were developed;
dedicated research teams were assembled; and formal means for
distributing results emerged. Each of these defining characteristics
deeply reflected the unique style of its respective community.

Although contributions to stellar astronomy were quite different in nature, individual views of the sidereal system not only
coincided closely but also achieved wide consensus among early
twentieth-century astronomers. Understanding the stellar universe

2

using statistical techniques, referred to as the "sidereal problem" in much of the technical literature, was considered the major (traditional) research program of stellar astronomy during much of the period considered here. In this study, we will deal primarily with developments only within those communities concerned with understanding the *cosmology* of the sidereal universe using statistical approaches. Although undertaking much valuable and even indispensable work, some groups, such as those led by Pickering, Hale, and Campbell in the United States and by Eddington and Dyson in England, did not fully contribute themselves to a complete solution to the central problem that came to define this research tradition. And numerous lesser-known astronomers contributed to various problems in stellar statistics, but did not themselves engage in statistical astronomy as cosmology. Rather, they worked on tertiary though important problems outlined by the recognized leaders. In the present study the emphasis throughout will be on statistical cosmology.

Because this is a historical study, I have employed a few terms describing people, events, issues, and ideas that were not in use at the time. For example, to refer to the Milky Way system as a "galaxy" during this period is an anachronism. Throughout I have generally deferred to the historical terms "stellar system," "stellar universe," or "sidereal universe" in referring to models of the Milky Way Galaxy. In contrast, although the terms "statistical astronomy," "cosmology," "statistical cosmology," and their derivatives were not in general use, I have used these terms with the understanding that the reader will find them descriptive and appropriate. Historically, the term statistical astronomy refers to observational studies of stars and galaxies – positions, motions, brightnesses, spectra, and so on – that rely fundamentally on statistical methodologies. This term therefore will be used when emphasizing technical problems relating to method or to a specific empirical problem. To the degree that astronomical studies were primarily motivated, particularly as in the case of Kapteyn and Seeliger, with understanding the macrofeatures of the Universe, the term statistical cosmology is appropriate. In order to emphasize that these latter research programs were fundamentally cosmological, this distinction will be used throughout this study.

Finally, this is a study in the development of early twentieth-century cosmology and not in the history of statistics and probability theory. Still, historically the two disciplines have been closely allied. Applied to astronomy, statistical methodology has a long and distinguished history. As recognized early in the nineteenth century by the Belgian statistician Adolphe Quetelet, the astronomer's love of natural order provided the foundation for statistical science:

> It is not to doctors that we owe the first tables of mortality, they were calculated by the celebrated astronomer [Edmond] Halley. . . . The laws that concern man, and those that govern social development, have always had a special attraction for the philosopher, and perhaps most especially for those who have directed their attention to the system of the universe.[2]

Like Quetelet before him, the nineteenth-century French statistician Antoine Augustin Cournot also took the celestial sciences as his model: "In a word, just as observational astronomy is the model for sciences of observation, and theoretical astronomy the model for scientific theories, so likewise should the statistics of stars ... serve some day as the model for all other statistical inquiries."[3]

With the possible exception of the Swedish statistical cosmologist Charlier, however, Seeliger, Kapteyn, and virtually all those who contributed to stellar statistics qua astronomy were astronomers rather than mathematicians or statisticians. To be sure, as already noted, these astronomers came from a long tradition of using statistics (and mathematics, of course) in the analysis of stellar data. But their aim was rarely in advancing the frontiers of statistics and mathematics qua methodology.[4]

The present study, which deals exclusively with stellar astronomy and how that research tradition shaped the dominant

[2] Quoted in Theodore M. Porter, *The Rise of Statistical Thinking, 1820–1900* (Princeton: Princeton University Press, 1986), p. 44.
[3] Ibid., p. 235.
[4] For an eighteenth-century attempt to assess certain astronomical problems using probabilistic reasoning, see Barry Gower, "Astronomy and Probability: Forbes versus Michell on the Distribution of the Stars," *Annals of Science,* 39 (1982), 145–60.

pre-relativistic cosmology, is divided into three major parts. Because the nascent statistical approach to astronomy examined stars in the aggregate, Part I examines methodological techniques specific to astronomy that came to be developed during the nineteenth century. The application of existing mathematical and statistical techniques during the period was, however, largely foreign to astronomical developments in this and most other fields. Still, important aspects of stellar statistics emerged, particularly after publication in 1859 of the first, fully comprehensive star catalogue, the *Bonner Durchmusterung*, which later came to play a crucial role in the rise of statistical cosmology.

During the nineteenth century, the most important issues dealt with the production of star catalogues, and achieving a basic understanding of the distribution of stars and the systematic stellar motions for developing a picture of cosmology. Here, a variety of technical and methodological issues prevail, including: Did Herschel's program for the "Construction of the Heavens," enunciated during the 1780s, actually lead to the development of a research program or tradition prior to the 1880s? If so, how did Herschel's program define and clarify critical research problems for nineteenth-century stellar astronomers? By the end of the century, what kinds of technical problems were considered central to this discipline? What sorts of methodological and ontological assumptions were crucial to statistical cosmology? Did these assumptions remain the same as various technical problems were examined? Was there consensus concerning the importance of such studies? And if so, when did it emerge, and from whom? Finally, if there was consensus, how strong was it, and was it accepted throughout the astronomical community?

The nineteenth-century background of these more recent developments has generally been neglected.[5] The principal concern of the first two chapters is to consider these earlier developments and focus on a set of assumptions, codified in William Herschel's

---

[5] For a brief discussion of some of the technical developments, see A. S. Eddington, *Stellar Movements and the Structure of the Universe* (London: Macmillan & Co., 1914), 184–231; some relevant historical remarks can be found in A. Pannekoek, *A History of Astronomy* (New York: Barnes and Noble, 1961), 444–7, 467, and S. L. Jaki, *The Milky Way: An Elusive Road for Science* (New York: Science History Publ., 1972), 221–89.

program, that persisted with little variation through most of the nineteenth century. Specifically, beginning with the work of William Herschel in the late eighteenth century, Chapter 1 explores the early nineteenth-century background in stellar studies focusing on those areas of research that provided the incentive for the emergence of a statistical study of the cosmos. Chapter 2 focuses on the emergence of two complementary themes that dominate stellar astronomy: the construction of the necessary star catalogues, and an increasing concern with the mathematico-statistical methods needed to analyze those catalogues. These chapters survey only those problems and themes that were needed by Seeliger, Kapteyn, and others for the prosecution of a statistical approach to cosmology.

By late nineteenth-century standards, Herschel's formulation of the problem became entirely inadequate. It was Seeliger, Kapteyn, and a few others, during the early years of the twentieth century, who eventually developed a methodology commensurate with the task at hand. Eventually, their statistical research evolved into an understanding of three basic stellar relationships: (1) How stars are actually distributed through space, (2) how their stellar luminosities vary from the intrinsically faintest to the very brightest, and (3) how stellar motions are distributed from the relative slowest to the fastest moving stars. These astronomers came to believe that a determination of the distribution, brightness, and velocity characteristics of stars as a whole would provide a full explanation of the sidereal universe.[6] Such relationships, in fact, had been Herschel's goal a century-and-a-half earlier. Even though statistical astronomy has since incorporated many additional topics,[7] at the time an understanding of these extremely complex relationships constituted the essential features of this branch of astronomy. During this period, this field of stellar astronomy increasingly held immense promise for an understanding of the distribution of the stars in space. As Seeliger, the earliest principal

---

[6] See, for instance, A. S. Eddington, "The Statistical Laws of Stellar Astronomy," *The Observatory*, 41 (1918), 384–6; and W. A. Schouten, *On the Determination of the Principal Laws of Statistical Astronomy* (Amsterdam: W. Kirchner Publ., 1918).

[7] See E. von der Pahlen, *Lehrbuch der Stellarstatistik* (Leipzig: Johann Ambrosius Barth Verlag, 1937); and R. J. Trumpler and H. F. Weaver, *Statistical Astronomy* (Berkeley: University of California Press, 1953).

pioneer of statistical cosmology, emphasized in 1904: "To every important problem in stellar astronomy . . . belongs the question of the spatial distribution of the stars."[8] The historical importance of statistical astronomy cannot be overestimated, for during the period from about 1900 to 1920, which delineates the bulk of the present study as Part II, statistical cosmology achieved its most enthusiastic level of support as a major research program in astronomy.

Although Herschel's assumptions informed stellar statistics at the most fundamental level, it was precisely their perceived inadequacy that eventually led (around the beginning years of the twentieth century) to a basic theoretical and methodological reorganization within stellar statistics. Chapters 3 and 4 respectively detail Seeliger's and Kapteyn's work and how they arrived at their views of the cosmos – empirically, conceptually, and methodologically. Seeliger and Kapteyn addressed nearly identical concerns, but because they worked largely independently of one another, the material for these chapters has been independently developed. Chapter 5 deals with the principal set of research problems that came to characterize statistical astronomy. Even though Seeliger and Kapteyn formulated most of these crucial problems, this chapter explores those concerns defined by this branch of astronomy as the important areas of research. The mature cosmologies of Seeliger and Kapteyn, those developed around 1920, are discussed in Chapter 6. Chapter 7 deals with the institutional and social context of statistical cosmology during the period from 1900 through about 1920 in order to provide an assessment of the degree and extent to which statistical astronomy was accepted among astronomers as the primary means for understanding the cosmos.

Part III is divided into two concluding chapters that explore the declining years of statistical cosmology as a major research program. Chapter 8 provides an analysis of the technical reaction by many of the younger astronomers within the astronomical community to the cosmologies of Kapteyn and Seeliger. The Conclusion (Chapter 9) briefly surveys the most significant

---

[8] H. von Seeliger, quoted in H. C. Vogel, ed., *Newcombe-Engelmann populäre Astronomie* (Leipzig: Wilhelm Engelmann Verlag, 3rd ed., 1905), p. 611.

developments that came to define the postclassical period of statistical cosmology. The period covered in these last two chapters, roughly 1918 to 1930, is frequently identified as the "golden age" or the "second astronomical revolution." Among the many scientists who have become most closely associated with the developments of the 1920s, Harlow Shapley, Edwin Hubble, Jan Oort, Bertil Lindblad, and Robert Trumpler particularly stand out. This was a time rich in ideas and radical in its consequences. Many of the central images that eventually came to be identified with twentieth-century cosmology – "the island-universe theory," "differential galactic rotation," the "expanding universe," "dark matter," and later the "big bang" theory – were born in the work of the astronomers and cosmologists of the 1920s.

This book begins with the emergence of statistical astronomy and ends with its demise – as a cosmology. In so doing, we will come full circle. Chapters 1 and 2 as background material, however, are not intended to be a comprehensive exploration of all statistically related problems in stellar astronomy; there are numerous historical problems during this period that will still provide grist for dissertations and research articles. Moreover, being a survey, these two chapters do not rely primarily on archival material. Chapters 3 through 8, however, present an analysis of issues and problems that characterize the mature period of statistical cosmology – roughly 1890 to about 1924. As such they are based heavily on archival and primary material gleaned from many libraries and research centers in Germany, Holland, Sweden, and the United States.

The present study, which has been in the making for many years, represents a sort of excavation of material dealing with statistical cosmology that has been buried for nearly a century. There is some very important material summarizing many of the technical issues written by several of the primary participants in statistical astronomy. With the exception of some of my earlier work there are only a few scholars who have begun to plow through an immense amount of material in stellar statistics – primary, archival, and scientifically technical – that must appear to the uninitiated as nothing less than intimidating. Although these sources have informed the debate concerning issues of the spiral nebulae, the island universe, and the expanding universe, all of which have been well mapped out, indirectly this study expands on the invaluable work of several scholars, particularly that of Professors Michael A. Hoskin and Robert W. Smith. Even though I have previously published on the nature of the Kapteyn Universe and on Shapley's reaction to Kapteyn's ideas, the present study has been extended enormously both in depth and breadth. The reader should understand that although the focus will almost exclusively be limited to the cosmology of statistical astronomy, there remain many historical issues still to be explored in the broader field of stellar statistics.

Part I

THE NINETEENTH-CENTURY BACKGROUND

I

EARLY NINETEENTH-CENTURY
STATISTICAL ASTRONOMY

Since antiquity the most obvious features of stars have been their positions and brightnesses. Astronomers and other celestial observers have identified a large number of stars, and organized their configurations into about 100 different constellations. Not until the eighteenth century, however, did astronomers examine stellar phenomena in the aggregate, and then only modestly. The systematic statistical (aggregate) basis to studying stellar phenomena originated only at the end of the eighteenth century with William Herschel, whose project of determining the position and spatial distribution of all stars, which he called the "Construction of the Heavens," constitutes his views of cosmology and cosmogony. A century later Kapteyn was to call it the "sidereal problem." Although the emergence of astrophysics during the latter half of the nineteenth century and continuing through the period of this study expanded into what is today considered normative astronomical research, it was still widely believed by many astronomers at the time that "among the problems of astronomy, the supreme place is held by that popularly known as the construction of the heavens."[1]

WILLIAM HERSCHEL AND THE "CONSTRUCTION OF THE HEAVENS"

William Herschel (1738–1822), a Hanoverian émigré and itinerant musician turned amateur astronomer, began his investigations of the sidereal system early in his astronomical career, eventually building the largest telescopes of his time. In his studies of sidereal

[1] H. MacPherson, Jr., "The Construction of the Heavens," *Popular Astronomy*, 14 (1906), 385 (385–93). Also on the importance of the sidereal problem, see the presidential address by Britain's Astronomer Royal, F. W. Dyson, "Construction of the Heavens," *Report of the British Association for the Advancement of Science*, (1915), 357–66.

astronomy, Herschel laid the groundwork for virtually all future investigations of the heavens. His work tended primarily in two directions: an understanding of the distribution of the stars in space, and discovery of the laws of stellar motion. Together, the solution to these problems was considered the principal aim in sidereal studies.

An understanding of the arrangement of the stars in space had already become a major concern in the eighteenth century when Herschel began his astronomical work. In the middle years of the century, the cosmologists Thomas Wright, Immanuel Kant, and Johann Lambert had begun to speculate on the nature of the stellar universe, and had proposed various (theoretical) cosmologies. Their ideas, all produced within a relatively short period from about 1750 to 1761, were based less on hard-core astronomical data, such as stellar positions, and more on a variety of speculative notions, especially the idea of a plurality of worlds. Their daring views extended well beyond a parochial concern with multiple worlds and theology, however, and suggested that the realm of stars and nebulae formed an interrelated whole.[2] In contrast to the more speculative approach of these cosmologists, the chief aim of astronomy, wrote Herschel, is to obtain "a knowledge of the construction of the heavens" that can be achieved by the determination of "the real place of every celestial object in space."[3] In order to determine the distribution of the stars, Herschel recognized the absolute need for accurately measured stellar distances. In this Herschel was half-a-century ahead of his time; not until the middle years of the nineteenth century did astronomers finally develop the techniques and instruments needed to obtain the required data. In the absence of such methods, Herschel suggested (1) that distance was proportional to stellar

[2] See Michael J. Crowe, *The Extraterrestrial Life Debate, 1750–1900: The Idea of a Plurality of Worlds from Kant to Lowell* (New York: Cambridge University Press, 1986), pp. 41–59.

[3] W. Herschel, "Astronomical Observations relating to the Construction of the Heavens, arranged for the Purposes of a Critical Examination, the Results of which appears to throw Some New Light upon the Organization of the Celestial Bodies," *Philosophical Transactions*, 101 (1811), 269 (269–336), and W. Herschel, "Astronomical Observations and Experiments tending to Investigate the Local Arrangement of the Celestial Bodies in Space," *Philosophical Transactions*, 107 (1817), 302 (302–31).

Figure 1   William Herschel. (Courtesy of Yerkes Observatory)

brightness by the principle of "faintness means farness." In this he assumed that stars are pretty much equally bright. In addition, Herschel assumed (2) that stars were more or less equally distributed throughout the sidereal system, and (3) that his telescopes could penetrate the edges of the sidereal system.[4] Using these assumptions, Herschel produced a three-dimensional model of the stellar universe in which the stars are distributed uniformly.

In a paper published in 1785, the year after Herschel began his examination of the sidereal system, Herschel first proposed his model of the stellar system using a method, called "star-gaging," that became basic to statistical cosmology.[5] This method entailed directly counting the total number of stars in a cone of space, vertex at the observer, in a particular direction. Herschel argued that he could observe all the stars in each direction (cone) to the edges of the stellar system, because his telescopes by assumption thoroughly penetrated the sidereal system. By the supposition that stars are equally distributed, Herschel calculated the distance to the stellar boundary, or, using Herschel's parlance, he was able to "gage" (gauge) the heavens.[6] Except for the brightest stars,

---

[4] These assumptions were not original with Herschel. Throughout his career, Herschel had been fascinated by Edmond Halley's model of the sidereal system (1720), which assumed that the stars are distributed more or less regularly throughout space. Michael Hoskin has shown that this assumption was Newton's in "Newton, Providence and the Universe of Stars," *Journal for the History of Astronomy*, 8 (1978), 77–101. Moreover, in 1747 James Bradley, who was interested in stellar parallax, encouraged astronomers to "examine nicely the relative situations of particular stars: and especially of those of the greatest lustre, which, it may be presumed lie nearest to us, and may therefore be subject to more sensible changes." See M. A. Hoskin, *William Herschel and the Construction of the Heavens* (London: Oldbourne Book Co., 1963), 27–42.

[5] W. Herschel, "Observations Tending to Investigate the Construction of the Heavens," *Philosophical Transactions*, 74 (1784), 437–51; and W. Herschel, "On the Construction of the Heavens," *Philosophical Transactions*, 75 (1785), 213–66.

[6] For additional details of Herschel's work not directly related to our purposes, see Hoskin, *William Herschel and the Construction of the Heavens*, 60–71, and S. Newcomb, *Popular Astronomy* (New York: Harper & Brothers Publ., 1882, 4th ed. revised), 477–86. On the technique of star-gauging, see R. A. Proctor, "Notes on Star-gauging," *Royal Astronomical Society, Monthly Notices*, 33 (1873), 461–4, and R. A. Proctor, "Further Notes on Star-gauging and on the Principles on which its Interpretation should depend," *Royal Astronomical Society, Monthly Notices*, 33 (1873), 535–6.

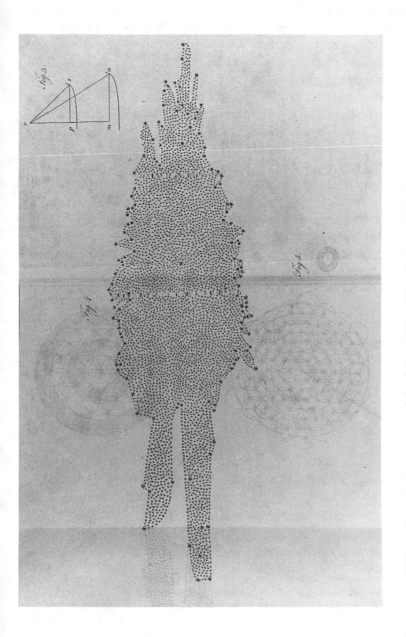

Figure 2   William Herschel's 1785 model of the Milky Way Galaxy. This diagram shows the Sun and the naked-eye stars in a cross-sectional plane of the Galaxy. (From Table VIII, Figure 4 of the *Philosophical Transactions of the Royal Society of London*, Part I, vol., 75, 1785)

Herschel noted the unusual concentration of all stars in the plane of the Milky Way, thus suggesting that the stars defined a highly flattened system (see Figure 2) whose diameter was large compared to its thickness. The Sun was nearly in the central plane of this system where the Milky Way, seen from the inside as the optical effect of our immersion in a stratum of stars, divides into two parts from Cygnus to Scorpio.

Herschel proposed not only a model in which the stars were assumed to be equally distributed and equally bright, and whose dimensions could be calculated, but, perhaps more important, he also devised a counting method in which stars were examined collectively, not individually. Eventually, Herschel realized that the stellar model that he had constructed (in 1785) was not altogether unproblematic. For example, he had built his (observational) cosmology on the principle of equal stellar brightness, even though he was aware, at least by the early years of the nineteenth century, of the contradiction implied by the existence of binary and multiple star systems with members of different luminosities. That is, he found star systems whose member stars were unequally bright, thus invalidating his supposition that stars are inherently equally bright and appear unequal only because they are to be assumed to be at unequal distances from the observer. Even his own work on double stars as early as the 1780s suggested that brightness did not directly represent distance.[7]

By 1817 Herschel recognized that his 1785 system, still considered by many scholars today as representing the quintessential Herschel, was astronomically untenable. In 1817 he wrote that his "star-gages, which on a supposition of an equality of scattering were looked upon as gages of distances, I have now to remark that, although a greater number of stars in the field of view is generally an indication of their greater distance from us, these gages, in fact, relate more immediately to the scattering of the

---

[7] Concerning the "brightness equals nearness" principle, see Hoskin, *William Herschel and the Construction of the Heavens*, 27–42. For an eighteenth-century attempt to address the question of multiple star clustering using probabilistic reasoning, see Barry Gower, "Astronomy and Probability: Forbes versus Michell on the Distribution of the Stars," *Annals of Science*, 39 (1982), 145–60.

stars. . . . ."[8] These views were already supported by numerous investigations of nebulae and the clustering of stars begun by Herschel as early as 1784.[9] And finally, he eventually realized that the twenty-foot telescope, which he had used in his 1785 investigations, "could not fathom the Profundity of the milky way."[10] Indeed, by 1817 he believed that even with the forty-foot telescope the Milky Way was "fathomless" and that "with this instrument [it] would carry the extent of this brilliant arrangement of stars as far into space as its penetrating power can reach . . . and that it would then probably leave us again in the same uncertainty as the 20 foot telescope."[11]

Without the uniform distribution postulate and the principle that brightness directly represents distance, it would not have been possible to construct a three-dimensional stellar model without accurately determined positional and intrinsic brightness data. Herschel understood clearly, however, that whereas it was one thing to realize that these assumptions were invalid, it was quite another to develop a model of the sidereal system without them. It is a testimony to Herschel's stellar program that,

[8] Herschel, "Astronomical Observations and Experiments tending to Investigate the Local Arrangement of the Celestial Bodies in Space," p. 325; also see Herschel, "Astronomical Observations relating to the Construction of the Heavens," pp. 269–336.

[9] On Herschel's "star-gauges," see his "Observations Tending to Investigate the Construction of the Heavens," pp. 437–51, and "On the Construction of the Heavens," pp. 213–66. For Herschel's major investigation of nebula and star clusters, see his "On Nebulous Stars, properly so called," *Philosophical Transactions*, 81 (1791) 71–88, "Astronomical Observations Relating to the Construction of the Heavens," pp. 269–336, and "Astronomical observations Relating to the Sidereal Part of the Heavens, and its Connection with the Nebulous Part: Arranged for the Purpose of a Critical Examination," *Philosophical Transactions*, 104 (1814), 248–84.

[10] Herschel, "Astronomical Observations relating to the Construction of the Heavens," p. 326.

[11] Ibid., p. 327. For a thorough discussion of this issue, see J. A. Bennett, "On the Power of Penetrating into Space: The Telescopes of William Herschel," *Journal for the History of Astronomy*, 7 (1976), 75–108. Indeed, as early as 1767, John Michell had already implicitly challenged the first two of these assumptions. See J. Michell, "An Inquiry into the Probable Parallax, and Magnitude of the Fixed Stars, from the Quantity of light which they afford us, and the Particular Circumstances of their Situation," *Philosophical Transactions*, 57 (1767), 234–64.

throughout most of the nineteenth century, these assumptions in various forms largely remained central to constructing coherent models of the sidereal system. At the threshold of the century, it was quite an achievement for Herschel, or for anyone, to construct an observational model of the stellar universe, and, in so doing, to demonstrate the possibilities latent in a statistical method. As a result, Herschel introduced a new methodological approach into the body of astronomical literature by demonstrating the inherent possibilities of counting stars.

F. G. W. STRUVE AND GALACTIC THEORY, CIRCA 1850

William Herschel's model of the stellar system and his numerous contributions to galactic research dominated astronomical thinking during the first third of the nineteenth century. Consequently, although some astronomers continued toward cosmological speculations in the tradition of Wright, Kant, and Lambert, through the 1820s astronomers increasingly focused on specific observational data, principally stellar brightness and magnitude, as an indication of stellar variety. This is not to suggest, however, that the determination of real stellar distances was no longer sought after. Indeed, Edmond Halley (1656–1742), the second Astronomer Royal of England, had already argued that stellar distances and stellar motions were very likely correlated.[12] Because it is nearly the first fact of observational astronomy that stars are distributed in three-dimensional space, sure knowledge of stellar motions had only begun with Halley's unexpected discovery in 1718 of the motion in latitude of various stars. In 1783, Herschel became the first astronomer to apply knowledge of proper motions in the derivation of the "solar apex," the direction toward which the Solar System is moving. Because Herschel had shown earlier that a star's apparent brightness is not a reliable measure of its distance, some astronomers were beginning to suggest that large proper motion could be used as a more reliable guide to the nearness of a star. Indeed, this principle underlay the campaign

[12] E. Halley, "Considerations on the Change of the Latitudes of Some of the Principal Fixt Stars," *Philosophical Transactions*, 30 (1717–19), 737 (736–8).

by the German F. W. Bessel (1784–1846), one of the foremost astronomers of the nineteenth century, to use the "flying star," 61 Cygni, a star with an unusually large proper motion, for the first successful measurement of trigonometric parallax. The discovery of stellar parallax provided not only additional confirmation for the Copernican (heliocentric) hypothesis, but it also provided the first direct measurement of stellar distances.[13] Because its importance could hardly be overestimated, the Royal Astronomical Society awarded Bessel its Gold Medal (his second) in 1841 for this work.[14]

Before Bessel's work all attempts at constructing a spatial arrangement of the stars relied on assumptions, such as the brightness-nearness principle, that contradicted various observational data. Rather than adopt complex, generally ad hoc assumptions, however, astronomers continued to rely on basic assumptions involving brightness, motion, and distance. They also recognized that, in order to examine how stars are distributed through space, star catalogues containing the magnitudes of large numbers of individual stars were needed.

Though such catalogues date from the time of the Greek philosopher Hipparchus in the second century B.C.E, during the early decades of the nineteenth century a variety of stellar catalogues were produced that provided the necessary astronomical information. Among the most important of these catalogues, which were eventually used in studies on the spatial distribution of the stars, were G. Piazzi's catalogue (1814), Bessel's reduction of J. Bradley's 3,000 stars (1818),[15] K. L. Harding's catalogue for

[13] In 1725, James Bradley discovered "stellar aberration," the first observational evidence for the Earth's motion about the Sun; see Pannekoek, *A History of Astronomy* (New York: Barnes and Noble, 1961), pp. 289–90.

[14] On the 61 Cygni episode, see M. A. Hoskin, *Stellar Astronomy: Historical Studies* (Bucks, England: Science History Publ., 1982), pp. 5–21 (p. 9); and A. H. Batten, *Resolute and Undertaking Characters: The Lives of Wilhelm and Otto Struve* (Dordrecht: D. Reidel Publ., 1988), pp. 113–29.

[15] A. M. Clerke, *A Popular History of Astronomy During the 19th Century*, (London: Adam and Charles Black Publ., 1908, 4th ed.), p. 32. For historical details concerning nineteenth-century star catalogues, see R. Grant, *History of Physical Astronomy* (1852), 506–14, S. Newcomb, *The Stars* (New York: G. P. Putnam's Sons, 1901), chap. IV, and Pannekoek, *History of Astronomy*, pp. 324–7.

asteroidal research (1822),[16] F. W. A. Argelander's naked-eye star atlas (1843),[17] and M. Weisse's reduction of 32,000 stars from Bessel's zone work (1846).[18] Before an understanding of the spatial distributions of stars could be accurately determined, precise knowledge of the positions of stars and their brightnesses was needed. It was widely agreed, beginning as we have seen with William Herschel himself, that basic assumptions about star brightness and stellar uniformity were highly problematic. Consequently, virtually all studies that examined questions of the distributions of stars began with a careful analysis and reduction of this basic data.

In 1847 Friedrich Georg Wilhelm Struve (1793–1864) of the Pulkovo Observatory in St. Petersburg published his analysis, the *Etudes d'astronomie stellaire*, on the spatial distribution of the stars.[19] Based on the Bessel–Weisse catalogue, which was limited to stars between ±15 degrees declination, and Argelander's work on the naked-eye stars, the *Etudes* provided the first major advance on the sidereal question after William Herschel's work a half-century earlier. By examining questions of stellar positions and magnitudes, rather than by considering macrofeatures of the stellar system such as its shape and relative size as Herschel had done earlier, Struve was able to determine certain quantitative relationships about the sidereal system. For example, whereas stars of all magnitudes were equally distributed along the equator, Struve discovered that the maximum density of stars tended to be asymmetrically distributed about the celestial equator. In

---

[16] K. L. Harding, *Atlas novus coelestis* (Göttingen, 1822). Also see J. C. Houzeau, "Uranometrie Generale, avec une Etude sur la Distribution des Etoiles visibles a l'Oeil Nu," *Bruxelles Observatoire Annals*, 1 (1878), 27 (26–37).

[17] F. W. A. Argelander, *Uranometria Nova* (1843).

[18] "Positiones mediae stellarum fixarum in zonis Regiomentanis a Besselio inter −15 et +15 declinationis observatarum, ad annum 1825 reductae et in catalogum ordinatae, auctore M. Weisse," *Jussu Academiae Imperiales Petropolitanae edi curavit et praefatus est F. G. W. Struve* (Petropoli, 1846).

[19] F. G. W. Struve, *Etudes d'astronomie stellaire sur la voie lactée et sur la distance des étoiles fixes* (St. Petersbourg, 1847). For a brief summary of this important work, see A. J. M. Szanser, "F. G. W. Struve (1793–1864). Astronomer at the Pulkovo Observatory," *Annals of Science*, 28 (1972), 338–43, (327–46), and most recently Batten, *The Lives of Wilhelm and Otto Struve*, pp. 145–55.

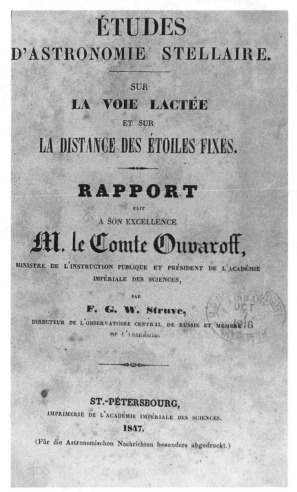

Figure 3   Title page of F. G. W. Struve's *Etudes d'astronomie stellaire* (1847).

Struve's opinion, this suggested that the Sun was not centrally located, but positioned eccentrically with respect to the equator and the Milky Way. He concluded that the true center of the sidereal system lay in a southerly galactic latitude.[20] He also confirmed Herschel's observation that the stars appear to be denser

[20]   Struve, *Etudes*, pp. 55–62.

near the Milky Way and increasingly sparse toward the galactic poles.[21]

Converting his data from celestial to galactic coordinates, Struve carefully examined Herschel's fundamental assumption of the uniform distribution of stars. The problem, as Struve made clear, was basically complicated by the fact that there were two mutually exclusive assumptions: Either stars have equal intrinsic brightness or they display a uniform distribution by brightness class. Which assumption, if either Struve asked, was correct? On the one hand, by assuming that stars are intrinsically equally bright, Struve showed that the stars should be uniformly distributed. On the other hand, assuming that a progression of stellar brightnesses from the first to the sixth magnitude forms a geometric series, Struve found that stellar density diminishes with distance.[22]

Attempting to reconcile these conflicting results, Struve wondered whether the added assumption of a light absorption in space could resolve the conflict. Although the idea of an absorbing medium in space, first proposed by the German physician and amateur astronomer H. W. M. Olbers (1758–1840) in 1823, would not only further complicate an already difficult situation, it could also be used to sort out some of the possibilities.[23] Olbers had argued that if space is isotropic and infinitely extended containing an infinite number of stars, then the sky should be ablaze with starlight at all times. To admit that space is unlimited, though, was to allow for the possibility of infinite space and the problem entailed in the optical paradox of an infinity of stars. Olbers provided no quantitative or empirical proof for his claim; his objection was strictly logical. On the other hand, Struve had some empirical data. To avoid Olbers's logical trap and to reconcile his conflicting findings, Struve became convinced of the existence of spatial absorption. He therefore derived an absorption

---

[21] Ibid., pp. 63–5, 71–4.   [22] Ibid., pp. 78–80.
[23] W. Olbers, "Ueber die Durchsichtigkeit des Weltraums," in J. E. Bode, ed., *Berliner Astronomisches Jahrbuch für das Jahr 1826* (Berlin, 1823), 110–21. For a discussion of Olbers's solution, see S. L. Jaki, *The Paradox of Olbers' Paradox: A Case History of Scientific Thought* (New York: Herder and Herder, 1969), pp. 131–43, 84–93.

coefficient so that "the intensity of light decreased in a greater proportion than the inverse square of the distances."[24]

The kernel of Struve's system is perhaps best summarized by one of Struve's harshest contemporary critics, the German astronomer J. E. Encke (1791–1865), director of the Berlin Observatory at the time. Encke identified the following assumptions (or conclusions, depending on one's perspective) that Struve had made: (1) Over the entire sky, stars are not spatially uniformly distributed; (2) (but) within the plane of the Milky Way, stars are roughly uniformly distributed; (3) the relative density of those stars outside the plane of the Milky Way decreases toward the galactic poles; (4) (therefore) the entire sidereal system is arranged in uniformly dense planes parallel to the Milky Way; and (5) the apparent brightness of a star is only roughly a measure of its distance.[25] Thus, of the three assumptions Herschel initially proposed, Struve questioned the validity of two of them.

Struve became the first astronomer within the research tradition Herschel outlined to attack the sidereal problem using primarily stellar counts. His work was most significant not only because he used the most reliable and up-to-date catalogue data, but also because his analysis demonstrated the possibilities inherent in stellar statistics that he so forcefully exploited along the lines Herschel suggested earlier. Struve's analysis produced interesting results, but he also recognized that his results were only as valid as a proper understanding of the data. Struve had examined the catalogue data carefully first, and had concluded that it was incomplete (especially at the higher magnitudes beyond eighth) because most stars had been observed only once, and, of course, because the data were limited to the relatively narrow, though critical, declination zone of ±15 degrees.

Virtually all subsequent developments in this nascent field of statistical cosmology have their roots in Struve's critique of Herschel's stellar assumptions. Although Encke was most critical, others, including George Biddell Airy (1801–92) in his 1847 presidential address to the Royal Astronomical Society, were highly

[24] Struve, *Etudes*, pp. 83–93.
[25] J. F. Encke, "Ueber die *Etudes d'astronomie stellaire*," *Astronomische Nachrichten*, 26 (1848), 337–50.

commendable of Struve. In the opinion of one of Struve's contemporaries, the *Etudes* "just issued from the press, treats of the distances and distributions of the fixed stars, presenting results of the highest importance in the future prosecution of Sidereal Astronomy."[26]

JOHN HERSCHEL AND MID-CENTURY COSMOLOGY

The most influential observational cosmologist of the middle years of the century was John Herschel (1792–1871), William's son and heir apparent. As with Struve, William Herschel's galactic assumptions provided discussion of much of the work in stellar studies through the nineteenth century. Just as Struve tried to relax Herschel's assumptions, so did John Herschel, though he adopted his father's view that the Milky Way is the optical effect of our immersion in a stratum of stars, also find much to criticize.

Still, John Herschel eventually rejected star gaging, the approach to stellar astronomy advocated by his father, and most others such as Struve, and increasingly focused on the study of nebulae as the key observational unit in cosmology. Thus while sweeping the northern sky, in due course Herschel examined the nebula M51 over the course of the two-year period from 1828 to 1830. After five lengthy observations (only three of which were useful, however), Herschel's observations with the eighteen-inch mirrors of his twenty-foot reflector led him to believe that the structure of this nebula was composed of a bright central region surrounded by a ring of stars. Herschel provided additional evidence to suggest that cosmologically M51 was composed of a primary circular ring at right angles to the line of sight together with a second semicircular ring inclined obliquely to and originating at opposite points with the primary ring.[27] Departing from the elder Herschel's cosmological views of the Milky Way, John Herschel used M51 as an analogous model of our Galaxy:

[26]   G. B. Airy, "Presidential Address," *Royal Astronomical Society, Monthly Notices,* 8 (1847–8), 91–5, and "Struve on the Distribution of the Stars," *Sidereal Messenger,* 2 (1848), 45 (45–7).
[27]   The importance of M51 in John Herschel's early work is described in M. A. Hoskin, "John Herschel's Cosmology," *Journal for the History of Astronomy,* 18(1) (1987), pp. 9–14 (1–34).

Figure 4   John Herschel, ca. 1850. (Courtesy of Yerkes Observatory)

Supposing it [M51] to consist of stars, the appearance it would present to a spectator placed on a planet attendant on one of them eccentrically situated toward the north preceding quarter of the central mass, would be exactly similar to that of the Milky Way, traversing in a manner precisely analogous [to] the firmament of large stars, into which the central cluster would be seen projected, and (owing to its greater distance) appearing, like it, to consist of stars much smaller than those in other parts of the heavens. Can it, then, be that we have here a brother-system bearing a real physical resemblance and strong analogy of structure to our own?[28]

This "central cluster and ring" model departed radically from his father's stratum model in which there were no gaps between the nearer (bright) stars and the more distant (fainter) stars. Thus although William's efforts to quantify his stratum model were mitigated ultimately by his stellar assumptions, John's theory of the Galaxy was far less dependent on his father's idea of star gaging and the distribution of the stars generally, and more crucially dependent on careful observations of nebulae.

In the year in which Struve published his *Etudes*, John Herschel published his *Results of Observations Made at the Cape of Good Hope*.[29] Although Herschel's cosmology was powerfully influenced initially by his father's theories and indeed survived in his popular *A Treatise on Astronomy* (1833), which was reprinted in numerous editions of *Outlines of Astronomy* (1849), the younger Herschel eventually broke radically from the ideas associated primarily with William. Herschel's *Cape Observations* (1847) was a compilation and distillation of extensive work undertaken in the southern hemisphere a decade earlier from 1834 to 1838, though a few years after Herschel developed his "central cluster and ring" model. The *Cape Observations* constitute the

[28] J. Herschel, "Observations of Nebulae and Clusters, Made at Slough, with a 20-feet Reflector, Between the Years 1825–1833," *Philosophical Transactions*, cxxiii (1833), 496–7 (359–506).

[29] John Herschel, *Results of Observations Made During the Years 1834, 5, 6, 7, 8 at the Cape of Good Hope; Being a Completion of a Telescopic Survey of the Whole Surface of the Visible Heavens, Commenced in 1825* (London: Smith, Elder & Co., 1847).

observational basis of John Herschel's mature cosmology, and they represent a sort of watershed in the younger Herschel's thinking.

John Herschel's observational work in nebular astronomy reached its zenith at the Cape, because not only was he able to undertake a "survey of the whole surface of the heavens" in order to obtain accurate position measurements of nebulae in the southern hemisphere to complement his father's work in the northern hemisphere, but also because the nebular studies urged a new cosmology even more complex than his earlier speculations on M51. In his observational work he realized the importance of accurately determining the magnitudes of stars. Although the elder Herschel had recognized the importance of quantitative compari sons of stellar brightness, it was not until John Herschel worked at the Cape that practical photometric work was introduced into the study of stellar astronomy.[30] In the process, Herschel's photometric techniques made his catalogues considerably more precise and therefore more valuable for investigations of stellar distributions than the earlier Bessel–Weisse zone catalogue used by Struve. Although Herschel possessed superior photometric knowledge necessary for the study of the distribution of stars, his interest primarily dealt with the study of nebulae, which was in keeping with the importance of his nebulae studies implied in his work on M51.

Because Herschel was an observational cosmologist dedicated to minute examination of the Milky Way, his conception of the structure of the Galaxy led him to believe it was immensely complex. For example, Herschel was the first to observe the general grouping of globular clusters in the direction of Sagittarius. He also noticed the difficulty of observing the Galaxy simply as another nebula, due to what is now known as the Zone of Avoidance. Although his southern observations reinforced his idea that the Galaxy "is not a mere stratum, but an annulus,"[31] as to the structure of the Galaxy itself, analogies with M51 disappeared altogether after 1834. His analysis of star counts in the *Cape Observations* demonstrated the differential distribution of stars:

[30] See Pannekoek, *History of Astronomy*, pp. 384–5.
[31] From Herschel's letter of June 1835 reprinted in M. A. Hoskin, "Astronomical Correspondence of William Rowan Hamilton," *Journal for the History of Astronomy*, 15 (1984), 69–73.

Whereas those brighter than eighth magnitude were uniformly distributed over the sky, stars of the ninth and tenth magnitudes increased with some frequency toward the Milky Way plane and stars yet more faint showed even more striking increase. By far the greater part of the Milky Way consists of these faintest stars.[32]

Even though his nebulae studies eventually forced him to break with his father's ideas, Herschel made some quantitative examinations of star distributions that suggested a variety of possibilities confirming and extending some of Struve's conclusions. For example, Herschel came to believe that (1) the brightest stars were nearest to the Sun, (2) the stars are more sparse and thin out far more quickly at right angles to the plane of the stratum (i.e., Milky Way), and (3) the Solar neighborhood is nearly empty.[33]

Together the work of William Herschel, Friedrich Struve, and John Herschel represents the most developed approach to early nineteenth-century stellar statistics, and ranks among the most innovative and potentially fruitful approaches to cosmology in general. Although the elder Herschel soundly demonstrated possibilities inherent in a statistical approach, Struve and the younger Herschel expanded this numerical understanding to the limits provided by the data available to them. In so doing, their methods and theories pressed the empirical data and scientific instruments to the point where further advances required wholly new approaches to organizing stellar phenomena.

[32] Herschel, "Of the Statistical Distribution of Stars," *Cape Observations*, chap. 4.1, art. 314.

[33] Herschel, *Cape Observations*, pp. 373–83.

2

STATISTICAL ASTRONOMY AND
THE MILKY WAY GALAXY

Through mid-century the various studies dealing with questions of the form and structure of the stellar universe were undertaken largely within the research program the elder Herschel had outlined. Given the problematic nature of the assumptions used and the incomplete stellar catalogues, however, these studies had reached only the most elementary quantitative conclusions. The most pressing need of this branch of astronomy, therefore, was two-fold: (1) The execution of a comprehensive inventory of stars in both the northern and the southern hemispheres involving accurate determinations of stellar magnitudes and positions, and (2) a thorough critique of Herschel's assumptions. During the 1850s, the first of these two problems was largely realized.

THE *BONNER DURCHMUSTERUNG* AND
STELLAR DISTRIBUTIONS

Bessel's earlier zone catalogue was designed primarily to facilitate the study of proper motions and the detection of stellar parallaxes. Therefore, it focused on star positions and not stellar magnitudes. For purposes of determining the spatial distributions of the stars, however, it was largely inadequate. Indeed, much of the more traditional study of the heavens focused on the stars as a mere backdrop to issues dealing with the planets and comets. Following in the tradition of the Herschels and of Struve, however, there slowly evolved a nascent interest in the spatial positioning of the stars. As a result, research of the stars themselves increasingly became a central focus. In order to fill this lacuna, beginning in 1852 Bessel's younger assistant, F. W. A. Argelander (1799–1875), director of the Bonn Observatory, compiled a comprehensive inventory of the telescopic stars between 6.0 and 9.5 magnitude. (Catalogues of the naked-eye stars were already available, including

Argelander's own 1843 star atlas.) Argelander, who wisely restricted himself to ninth magnitude stars, primarily because the
inclusion of still-higher magnitudes would have increased geometrically the numbers of stars and therefore extended completion of his project by years if not decades, finished his project of
surveying the entire northern hemisphere by 1859.[1] Argelander
compiled this data, containing the positions (in right ascension)
and magnitude estimates of about 324,000 stars, into a three-
volume catalogue called the *Bonner Durchmusterung des
Nordlichen Himmels*. Because of its comprehensiveness and reliability, it immediately eclipsed all other star surveys of the northern
hemisphere, and became an indispensable tool in many aspects of
astronomical work including studies in stellar distributions during the latter half of the nineteenth and early twentieth centuries.

To enhance the reliability of his observations, Argelander fixed
the position of each star in the *Bonner Durchmusterung*, as it
became known, twice. In addition to recording large numbers of
stars, Argelander developed a technique to guarantee the relative
consistency of the magnitude determinations of each of the stars.
Here Argelander assumed that a star of any magnitude, say the
seventh, would appear 2.519 times brighter than a star of the
next, or eighth, magnitude. Therefore he used what is called a
light-ratio equivalent to 2.519. Although there was little universal agreement on a standard light-ratio (even though virtually all
astronomers used pretty much the same value), Argelander's use
of the same, undeviating value throughout the entire *Bonner
Durchmusterung* strengthened its reliability. Since antiquity the
naked-eye stars (those brighter than apparent magnitude 6.0)
had been divided into six classes for reasons originating in
the physiology of the human eye. During the 1690s, Newton had
developed a model of the Universe in which stars were both
approximately equally distributed throughout and subject to
universal gravitational attraction.[2] In his model the stars, which

---

[1] See F. W. A. Argelander, *Bonner Durchmusterung des Nordlichen Himmels*
(Bonn: A. Marcus and E. Weber Verlag, 1903), 3 vols., preface; also see A.
Pannekoek, *A History of Astronomy* (New York: Barnes and Noble, 1961),
pp. 467–8.

[2] M. Hoskin, "Newton, Providence and the Universe of Stars," *Journal for the
History of Astronomy*, 8 (1978), 81–9.

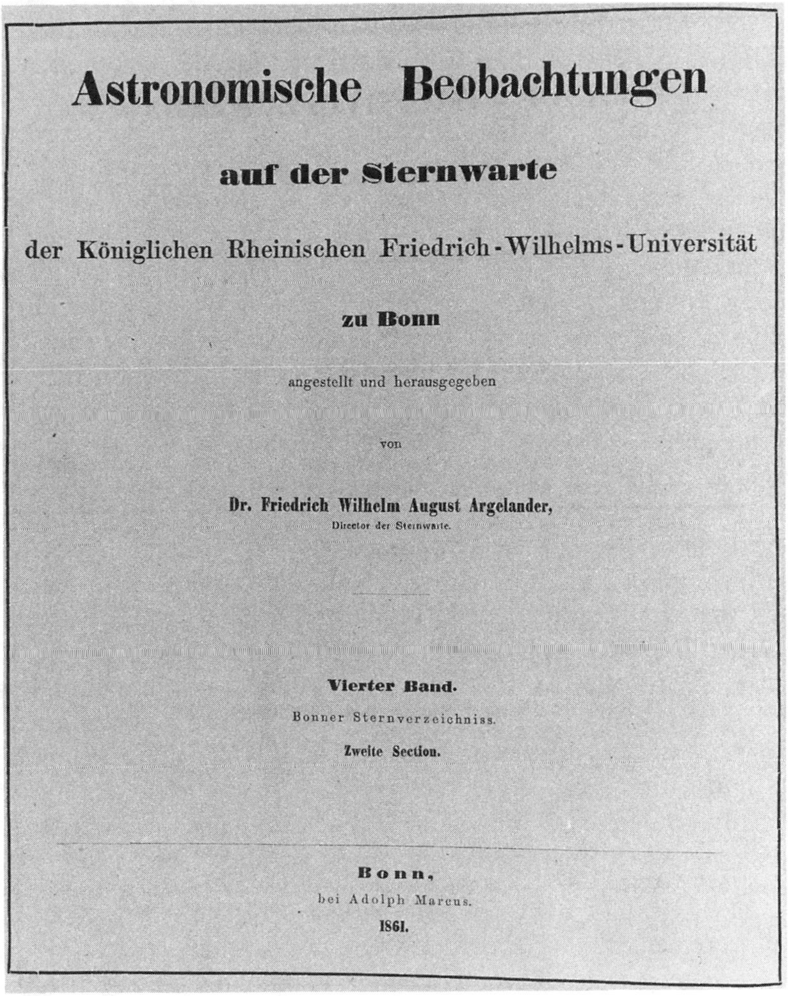

Astronomische Beobachtungen

auf der Sternwarte

der Königlichen Rheinischen Friedrich - Wilhelms - Universität

zu Bonn

angestellt und herausgegeben

von

Dr. Friedrich Wilhelm August Argelander,
Director der Sternwarte.

Vierter Band.

Bonner Sternverzeichniss.

Zweite Section.

Bonn,
bei Adolph Marcus.
1861.

Figure 5   Title page of F. W. A. Argelander's *Bonner Durchmusterung*.

were assumed to be uniformly bright, were grouped at distances corresponding to equally placed concentric spherical shells. In order to harmonize his model with the actual stellar data available at the time, Newton suggested that if first-magnitude stars correspond to one unit of distance, then sixth-magnitude stars correspond to a distance of eight or nine units. As it turns out, Newton's distance for sixth-magnitude stars is very close to the modern relationship between first and sixth magnitudes adopted after about 1860. Thus each successive magnitude class appears to be about two-and-a-half times fainter than the immediately preceding class. In the words of one contemporary historian of astronomy, his values, however, are "the result of luck rather than judgement."[3]

Knowledge provided by Newton's model of the Universe was largely indirect.[4] What is significant here is not whether direct influence can or cannot be established, but rather the fact that the problems inherent in defining a workable light-ratio between magnitude classes had already been considered in detail in the seventeenth century. Newton's contemporaries, the astronomer David Gregory and the cleric Richard Bentley, were very familiar with this part of Newton's work, and had exposed it to public consideration by the early decades of the eighteenth century. Bentley himself had first drawn Newton's attention to certain questions that culminated in Newton's model of the Universe. William Herschel and later his son, John, as well as Friedrich Struve all recognized the importance of describing a precise mathematical relationship of light-ratios for stellar magnitudes.

In 1856 the English astronomer Norman R. Pogson investigated the light-ratios used in prominent star catalogues, including the partially completed *Bonner Durchmusterung*, in order to standardize this increasingly essential, observational information. "The ratios thus found," wrote Pogson,

> were remarkably accordant: the mean of all, 2.4. . . . I selected 2.512 for convenience of calculation, as the reciprocal of $^1/_2 \log(r)$ [$r = 2.512$], a constant continually occurring in photometric formulae, is in this case exactly 5. . . . It appears, then, that whatever they may have intended, nearly all the great catalogue-makers have involuntarily fallen into this ratio, of one so nearly the same as to be practically identical.[5]

[3] Ibid., p. 89.   [4] Ibid., pp. 91–3.
[5] N. R. Pogson, "Magnitudes of Thirty-six of the Minor Planets for the first day of each month of the year 1857," *Royal Astronomical Society, Monthly*

Because the difference in brightness across five magnitudes is exactly 100 (i.e., 2.512^5), under his scheme Pogson defined the difference between two adjacent magnitude classes as 2.512. Thus he further confirmed the reliability of Argelander's approximate ratio (2.519).[6] Pogson's light-ratio came into general acceptance only slowly, however, and was not explicitly accepted until after about 1880 with the publication of the Harvard Photometry, the *Uranometria nova Oxoniensis*, and the Potsdam Photometric Survey.[7]

It was universally recognized that the importance of the *Bonner Durchmusterung* lay in the fact that it combined a consistent light-ratio with the completeness of star positions down to the ninth magnitude. Within the first decade of its publication, several studies of the Milky Way and the general distribution of the stars appeared all based on the *Bonner* stars. The earliest discussion to appear was by the German Naturalist Alexander von Humboldt, who, along with Charles Darwin, was one of the two most widely read interpreters of geographical exploration. Raising his sights, Humboldt led his readers into celestial regions of exploration in the 1861 edition of his *Cosmos*. Humboldt reiterated several questions of general interest in determining the "structure of the universe": How many stars are visible to the unaided eye?; how many of these have been catalogued and what

Notices, 17 (1856–7), 14 (12–15); Pogson's light-ratio had been partially anticipated by C. A. Steinheil, "Steinheil's photometer," *Poggendorff Annalen der Physik und Chemie*, 34 (1835), 644–51. Pogson's comment that "nearly all the great catalogue-makers have involuntarily fallen into this ratio ..." can be explained by the likelihood that they were fully aware of the problems Newton raised, and that their use of the general ratio 2.4 was neither coincidental nor involuntary.

6 Pogson, "Magnitudes of Thirty-six of the Minor Planets," pp. 14–15.

7 D. B. Herrmann, "N. R. Pogson and the Definition of the Astrophotometric Scale," *Journal of the British Astronomical Association*, 87 (1977), 146–9. For contemporary discussions of the psychological limits of this light-ratio, see G. T. Fechner, "Ueber ein wichtiges psychophysisches Gesetz und dessen Beziehung zur Schatzung der Sterngrossen," *Abhandlungen der Mathematisch-Physischen Klasse der K. Sachsischen Gesellschaft der Wissenschaft zu Leipzig*, 4 (1859), 455ff., and G. T. Fechner, "Ueber die Frage des psychophysischen Grundgesetzes mit Rucksicht auf Aubert's Versuche," *Berichte über die Verhandlungen der Mathematisch-Physischen Klasse der K. Sachsischen Gesellschaft der Wissenschaft zu Leipzig*, 16 (1864), 1–20.

are their positions?; and what is the progression of their magnitudes? Others followed in a similar vein.[8]

Perhaps the earliest technical paper analyzing stellar distributions and luminosities using the *Bonner Durchmusterung* was published by the Austrian astronomer Karl L. von Littrow (1811–77) in 1869–70.[9] Even though, as we have repeatedly seen, William Herschel's basic assumptions were technically unreliable, Littrow, using the newer *Bonner* data, wanted to test Herschel's two assertions, namely, that all stars possess the same absolute (or intrinsic) brightness and that stars are very nearly equally distributed throughout space, that is, the principle that "faintness means farness." Assuming all stars are equally bright and equally distributed, Littrow argued on theoretical grounds that we should expect the "brightness ratio," the numbers of stars per unit volume of space, to be a constant. To test this hypothesis, Littrow proposed, as Newton had done earlier, that all stars of a certain magnitude "m," say, are located between the distances r_m and r_{m+1} from the Sun.[10] That is, he partitioned space into concentric spherical shells about the Sun, so that the volume of any shell would be proportional to the number of stars corresponding to the magnitude class represented by the given shell. Beginning with the first magnitude class in the *Bonner Durchmusterung*, he

8   A. von Humboldt, *Cosmos: A Sketch of a Physical Description of the Universe* (New York: Harper and Brothers, 1861), vol. 3, pp. 103–8. See K. L. von Littrow, "Zahlung der Nordlichen Sterne im Bonner Sternverzeichnisse nach Grossen," *Astronomische Nachrichten*, 62 (1864), 357–62, and 73 (1869), 201–6.

9   K. L. von Littrow, "Zur Zahlung der nordlichen Sterne im Bonner Sternverzeichnisse nach Grossen," *Sitzungsberichte der Mathematisch-Physikalischen Klasse der K. Akademie der Wissenschaften in Wien*, 59(2), (1869), 569–96; and the related supplement "Nachtrag zu der Abhandlung: 'Zahlung der nordlichen Sterne im Bonner Verzeichnisse nach Grossen'," *Sitzungsberichte der Mathematisch-Physikalischen Klasse der K. Akademie der Wissenschaften in Wien*, 61 (Abt. 2) (1870), 263–6. These two papers must be read together, because the second paper corrects a number of mistakes made in the first.

10 Littrow first analyzed the *Bonner* star-counts for full magnitude differentials and then for half-magnitude differentials in order to enhance the credibility of his work; see Littrow, "Zur Zahlung der nordlichen Sterne," pp. 587–91 and 592–5, respectively. Struve had also used a similar partitioning scheme in his studies on the distributions of stars; see F. G. W. Struve, *Etudes d'astronomie stellaire sur la voie lactée et sur la distance des étoiles fixes* (St. Petersbourg, 1847), p. 79.

found that the star-counts in each class increased geometrically, which was to be expected, because the volumes of successive shells also increased in the same proportion. Finally, using the brightness corollary, Littrow calculated the theoretical number of stars that one would expect in each concentric shell. These theoretical values closely approximated the actual number of stars in each magnitude class. Littrow concluded jubilantly: "Throughout we have maintained the suppositions with which we began; the mutual equal distance and the equal brightnesses of stars is confirmed."[11]

Even though the *Bonner* stars were limited to the ninth magnitude class, Littrow did not hesitate to speculate on the total number of stars of all magnitudes (and, hence, the size of the Milky Way Galaxy) in the entire stellar system. Since antiquity it was well known that the fainter (higher) the magnitude class, the more stars each successive class contains and, therefore, the larger the system. Although telescopes in Littrow's day could fathom stars barely to the sixteenth (apparent) magnitude, Littrow estimated on the basis of this meager evidence that at most the entire stellar system contains around 1,500 million stars. On this last point, Littrow entirely begs the issue. The question of immediate and direct bearing concerns the real limiting magnitude (in absolute terms) of the least luminous stellar objects. If the Universe contains only a finite number of stars, then the optical and gravitational paradox entailed in an infinitely populated stellar universe could be avoided. Because there was no discrepancy in Littrow's system between the theoretical and actual star-counts, Littrow neither required the added assumption of an absorbing interstellar medium nor faced the consequences of Olbers's paradox. Despite Littrow's work, these cosmological questions had plagued astronomers at least since Newton, and they were again to receive renewed serious attention by the end of the century.[12]

Because of the completeness of the *Bonner Durchmusterung* down to the ninth magnitude, Littrow's analysis of the distribution

[11] Littrow, "Zur Zahlung der nordlichen Sterne," p. 591.
[12] See H. von Seeliger, "Ueber das Newton'sche Gravitationsgesetz," *Astronomische Nachrichten*, 137 (1895), 129–36, and C. V. L. Charlier, "Wie eine unendliche Welt aufgebaut sein kann," *Meddelanden Fran Lunds Astronomiska Observatorium* (Serie I), no. 38 (1908), 15 pp.

of the stars raised the discussion to a new level of sophistication. For the next decade Littrow's work provided astronomers with sufficient grist for their calculating mills. In 1872 the Swedish astronomer Hugo Gyldén (1841–96), who in 1889 would become president of the prestigious Deutsche Astronomische Gesellschaft (one of the few societies that maintained an international profile), provided the first critique of Littrow's work.[13] Although Gyldén accepted as a working supposition Littrow's assumption that the "hypothesis of the uniform distribution of stars in space is not altogether different from the real situation," he rejected, for reasons similar to those given by the elder Herschel and John Michell, among others, Littrow's additional assertion that all stars are intrinsically of equal brightness.[14] Instead, Gyldén suggested that the observational evidence on stellar brightness could be more convincingly preserved given the far more likely proposition that the intrinsic stellar brightness of all stars is, in some mathematical sense, normally distributed. In other words, all possible brightnesses, from the very faint to the very bright, could be represented by some frequency function, such as a Gaussian distribution. Gyldén used an integral expression to express his frequency function. Although he was one of the two foremost celestial mechanists of the latter half of the nineteenth century, he was unable fully to solve his integral equations.[15] Even though special solutions to such equations had been proposed by a few mathematicians during the nineteenth century, the general theory of integral equations was not formalized until 1896–97.[16] Therefore, although Gyldén's suggestion may have been very appealing, in 1872 very little could have been gained in pursuing

[13] H. Gyldén, "Relationer emellan stjernornas glans, antal och relativa medelafstand fran var standpunkt i verldsrymden," *Ofversigt af Kongl. Vetenskaps-Akademiens Fordandlingar. Stockholm,* 29 (7) (1872), 27–36. For a translation and discussion of this important paper, see my "An Early View of Galactic Structure: Hugo Gyldén and the `Fundamental Equation'," (unpublished ms.).

[14] H. Gyldén, "Relationer emallen stjernornas glans," p. 36.

[15] The other celestial mechanist was the Frenchman F. F. Tisserand (1845–96). These two astronomers, who died within a few days of one another, were the most prominent representatives of mathematical astronomy at the time of their deaths.

[16] M. Kline, *Mathematical Thought from Ancient to Modern Times* (New York: Oxford University Press, 1972), pp. 1052–75.

Figure 6   Hugo Gyldén, ca. 1885.

his idea of using an analytical expression for the luminosity function. A generation later, in the hands of Seeliger and Kapteyn, Gyldén's suggestion first proved to be the key idea that unlocked the full dimensions of stellar statistics.

Aside from the technical detail of requiring a general theory for such equations, were there other reasons why Gyldén's idea remained stillborn for several more decades? There appeared to be no rigorous method capable of utilizing his frequency function in an analysis of stellar brightnesses, but it is most likely that Gyldén's paper, which was written in Swedish and was the only one he was to write on this particular subject, remained hidden from the non-Swedish-speaking community. The earliest reference to Gyldén's paper does not appear until 1912,[17] and then only by another Swedish astronomer. It therefore appears likely that Gyldén's paper remained largely unknown because it had been written in his native language, in a Swedish journal for a Swedish reading audience. In contrast, whenever Gyldén wished to communicate with the larger astronomical community, he always wrote in German or French and published in major European scientific journals. Gyldén's ideas were pregnant with great possibilities, but for more than twenty-five years the lack of adequate mathematical and conceptual techniques prevented stellar astronomers from resolving the assumption of the equality of intrinsic stellar brightnesses.

Littrow's studies were also carefully examined by the American astronomer B. A. Gould (1824–96) in an 1874 paper.[18] Because Littrow's analysis is based exclusively on stars down to the limiting *Bonner* (ninth) magnitude, Gould tested Littrow's conclusions on empirical data that used stars with magnitudes that lay outside Littrow's range. Using both Eduard Heis's stellar catalogue, as well as his own data for the southern hemisphere, Gould confirmed Littrow's conclusions for stars brighter than the sixth

---

[17] C. V. L. Charlier, "Studies in Stellar Statistics: Constitution of the Milky Way," *Meddelanden Fran Lunds Astronomiska Observatorium* (Serie II), no. 8 (1912), 7. Gyldén published several very suggestive papers in obscure Swedish journals. His 1871 paper on galactic rotation was also initially published in Swedish; see V. E. Thoren, et al., "An Early View of Galactic Rotation," *Centaurus*, 18 (1974), 301–14.

[18] B. A. Gould, "On the Number and Distribution of the Bright Fixed Stars," *American Journal of Science and Art*, 8 (3rd series) (1874), 325–34.

magnitude, but persuasively argued that Littrow had oversimpli-
fied the problem for fainter magnitudes.[19]

Although Gould and a few other astronomers were sympathetic
to Littrow's approach,[20] it appears most astronomers rejected
Littrow's model because the two fundamental assumptions upon
which it was based were in conflict with other independently
obtained empirical results.[21] By the middle of the nineteenth century, double star astronomy had been advanced by many astronomers, including F. G. W. Struve, his son, Otto Struve, Bessel,
and John Herschel.[22] Already in the work of Michell and certainly in that of the elder Herschel, the equal brightness assumption was fundamentally challenged by the observation that double
and multiple star systems contain stars whose individual members have magnitudes that differ considerably from one another.
Furthermore, using direct, empirical data it was well understood
by 1875 that stars whose distances are known through parallax
measurement have magnitudes that do not correspond to magnitudes predicted by Littrow's assumptions.[23] Finally, as summarized by the English astronomer Richard A. Proctor, a long-
time antagonist of Herschel's assumptions, Littrow's method of
analysis obscured important details:

> Von Littrow has shown that when only the numerical
> relations of stars of various orders of brightness are
> considered, the increase in the number of fainter stars
> corresponds with the theory that apparent magnitude
> depends chiefly on distance. When, however, the local-
> ization of stars of various orders is considered, we find

[19] Ibid., pp. 333–4.
[20] See R. Falb, "Anzahl und Vertheilung der hellen Fixsterne," *Serius: Zeitschrift für populäre Astronomie*, 8 (1875), 49–52; and J. I. Plummer, "On the Collective Light and Distribution of the Fixed Stars," *Royal Astronomical Society, Monthly Notices*, 37 (1877), 436–9.
[21] For a contemporary critique of the proposition that stellar distances are directly related to stellar brightness, see E. J. Stone, "On Apparent Bright-ness as an Indication of Distance in Stellar Masses," *Royal Astronomical Society, Monthly Notices*, 37 (1877), 232–7.
[22] For the state of double star astronomy in 1850, see O. W. Struve, "Cata-logue revu et corrige des Etoiles doubles et multiples Decouvertes a l'observatoire Central de Poulkova," *Memoires de l'Academie Imperiale des Sciences de St. Petersbourg*, 7 (6th series) (1850), 385–405.
[23] See C. R. M., "Stellar Distribution – I," *English Mechanic*, 22 (1875), 291–2.

evidence of the numerical superabundance of stars really smaller than their neighbors.[24]

The research tradition outlined by William Herschel nearly a century earlier remained problematic through most of the nineteenth century. Although various stellar assumptions continued in use, the most significant developments were the production of the great survey catalogues, such as the *Bonner Durchmusterung*, because they provided the raw, empirical material needed to verify or dispute the assumptions on which the various models were based. Consequently, astronomers increasingly focused on the presumed accuracy of the star catalogues, particularly the *Bonner Durchmusterung*. By 1880, it was widely believed that prior to continued investigation of the distribution of the stars, it would be necessary to verify the raw, empirical data itself given in the star catalogues.[25]

COSMOLOGY AND STELLAR STUDIES, CIRCA 1880

Precise stellar magnitude determinations required careful comparisons with a standard source of light. Working at the Cape, John Herschel become the first astronomer to provide photometric comparisons of stellar magnitudes. Herschel's photometric

[24] R. A. Proctor, "Note on the Distribution of the Fixed Stars," *Royal Astronomical Society, Monthly Notices*, 37 (1877), 424 (424–5). Concerning the validity of Herschel's hypothesis, Proctor wrote in 1870: "I am fully sensible that it is only in certain popular treatises of astronomy that a belief in any thing like a real uniformity of structure in the sidereal system is attributed to astronomers in authority." See R. A. Proctor, *Other Worlds than Ours* (New York: P. F. Collier & Son, 1870), p. 258. Proctor explicitly examines and rejects the principle of the homogeneity of stellar distribution in "On a Chart of 324,128 Stars" (1870) and in "The Construction of the Heavens" (1871), both reprinted in *The Universe and the Coming Transits* (London: Longmans, Green & Co., 1874), 157–66 and 172–205.

[25] Before this issue was fully resolved by the end of the century, spirited debate developed; see J. I. Plummer, "On the Collective Light and Distribution of the Fixed Stars," *Royal Astronomical Society, Monthly Notices*, 37 (1877), 438–9; J. L. E. Dreyer, "Distribution of the Fixed Stars," *The Observatory*, 1 (7) (1877–8), 216; E. Schönfeld, "Berichte über die Thätigkeit der Sternwarten in Bonn," *Vierteljahrsschrift der Astronomische Gesellschaft*, 12 (1877), 41–55 (p. 47); J. I. Plummer, "Distribution of the Fixed Stars," *The Observatory*, 1 (8) (1877–8), 252; and E. Schönfeld, "Distribution of the Fixed Stars," *The Observatory*, 1 (9) (1878), 285 (284–5).

measurements, however, used a pointlike image of the moon for the standard of comparison, which, due to the unevenness and changing conditions of observation of the moon, introduced a systematic source of error. In 1861, the German astronomer J. C. F. Zöllner invented the first astrophotometer that relied entirely on an artificial source for magnitude comparisons, thus providing a mechanism for improving the accuracy of the magnitude estimates. Finding a way to apply these ideas to the great *Bonner* star catalogues was the work of the American Edward C. Pickering (1846–1919), director of the Harvard College Observatory.

In 1884 Pickering completed his "Photometric Catalogue" containing 4,260 stars brighter than the sixth magnitude. Because of its reliable magnitude determinations and the internal consistency of the data, Pickering's catalogue quickly became the standard reference work for sidereal studies of the naked-eye stars. Although a photometric revision of each of the 324,000 *Bonner* stars was an impractical task, Pickering devised a technique that enabled each of the *Bonner* magnitudes to be reduced to the photometric scale. Published in 1890, Pickering's "Photometric Revision of the Durchmusterung" provided accurate magnitude values of the stars between the sixth and ninth magnitudes. In addition to his 1884 revision, these catalogues assured astronomers that the magnitudes of all stars brighter than the ninth magnitude could be known with complete accuracy.[26]

Although Pickering in 1885 briefly described the geometric progression needed to predict the number of naked-eye stars found in each successive magnitude class of his "Photometric" catalogue,[27] it was the Italian astronomer Giovanni V. Schiaparelli (1835–1910) who first developed a theoretical model of stellar numbers and compared it with the actual star-counts found in Pickering's naked-eye catalogue.[28] Schiaparelli approached his

[26] E. C. Pickering, "Harvard Photometry," *Harvard College Observatory, Annals*, 14, part II (1885), 329–512; and E. C. Pickering, "Photometric Revision of the Durchmusterung," *Harvard College Observatory, Annals*, 24 (1890), 1–2 (1–199).

[27] E. C. Pickering, "Distribution of the Stars," *Harvard College Observatory, Annals*, 14 (2) (1885), 483–4 (477–84). For the stars in the southern hemisphere, Pickering relied on his photometric correction of Gould's catalogue.

[28] G. V. Schiaparelli, "Sulla Distribuzione Apparente delle Stelle Visibili ad Occhio Nudo," *Dalle Pubblicazioni del Reale Observatorio di Brera*

studies of the stellar phenomena within the traditional program, but he introduced one significant innovation: He represented the stellar assumptions mathematically, which, over the course of a century, had come to characterize studies of the spatial distributions of the stars.

Following the earlier work of Littrow, Schiaparelli assumed the existence of a stellar system of nearly unlimited extent with no spatial absorption in which stars are scattered uniformly with (constant) stellar brightnesses independent of the distance from the Sun. On the basis of these assumptions and using Pogson's light-ratio, Schiaparelli argued on theoretical grounds that the numbers of stars contained in successive spherical shells should increase by a factor roughly of 3.981. This means that one should expect the star-counts in successively higher magnitude classes nearly to quadruple. This increase in the numbers of stars in successive shells has been variously called the density-ratio, the star-ratio, and, as christened by the English cosmologist Arthur S. Eddington in 1914, the "fundamental theorem."[29]

Having devised this theoretical model, Schiaparelli, as Littrow and others had done earlier, followed the conventional pattern and examined the data in Pickering's "Photometric" catalogue of the naked-eye stars. He concluded that the number of stars in each magnitude class fails to increase according to the predictions provided by the model. In order to reconcile his theory with the data, Schiaparelli tried several different ideas including (1) using different initial density-ratios to describe the progression of stars in each magnitude class, and (2) invoking spatial light absorption to account for fewer stars within each class. None of these ad hoc approaches, however, mitigated sufficiently the difference between the predicted and the actual values found within the six magnitude classes of the "Photometric" catalogue, and therefore they raised only additional problems. For instance, the absorption coefficient Schiaparelli used suggested a density-ratio of 1.35 when extrapolated to the ninth magnitude. If true, this meant that there should be only a slight increase in the numbers of stars

in Milano, 34 (1889), 28 pp; reprinted in Le Opere di G. V. Schiaparelli (Johnson Reprint, Sources of Science – no. 69, 1968), vol. 7, 13–47.

[29]   A. S. Eddington, Stellar Movements and the Structure of the Universe (London: Macmillan & Co., 1914), pp. 184–5.

in the ninth class over the eighth. Such results, he concluded, were inadmissible.

Though Schiaparelli was ultimately unsuccessful, his work suggested the growing realization among stellar astronomers that new methods were required to solve the enigma of the stellar arrangement. The relatively little information available exclusively from star-counts had attained the necessary accuracy required for exact investigations only since Pickering's work. But the complexity of the problem involved in understanding the architecture of the stellar system remained. As one of the early users to incorporate the advantages of the superior Harvard catalogue, Schiaparelli's methods hinted at the direction in which statistical studies were to proceed.

PROPER MOTIONS, PARALLAXES, AND STELLAR DISTANCES

Studies in the distribution of the stars was only one of two main directions that statistical astronomy was heading. The other branch of this subdiscipline of astronomy dealt with the motions of stars. Knowledge of stellar motions had already begun with Edmond Halley's discovery in 1718 of the motion in latitude of three stars. In 1783, Herschel became the first to use known proper motions to derive the "solar apex," the direction toward which the Solar System is moving. By this time, some astronomers were beginning to suspect that large proper motion might be a more reliable guide to the nearness of a star than apparent brightness, and this principle underlay Bessel's efforts to use 61 Cygni for the first successful measurement of trigonometric parallax.[30] Later in the century astronomers such as Argelander and Airy, both of whom used Gauss's least-squares techniques to identify random fluctuations in stellar motions, eventually concluded that random irregularities were due, not to observational errors, but rather to the *peculiar* or individual motions of stars relative to other stars.[31]

---

[30] See M. A. Hoskin, *Stellar Astronomy: Historical Studies* (Bucks, England: Science History Publ., 1982), pp. 5–21, particularly p. 9.

[31] F. W. A. Argelander, "Ueber die eigene Bewegung des Sonnensystems," *Astronomische Nachrichten*, 16 (1839), 43–56, and G. B. Airy, "On the Movement of the Solar Systems," *Royal Astronomical Society, Memoirs*, 28 (1860), 143–71.

They recognized the systematic nature of proper motion data, and emphasized the need to reduce the data to yield peculiar motions. As a result, during the last half of the century there came into widespread use the assumption of randomness among the real motions of stars.

This assumption was coupled with the belief that proper motion data could be used as a measure of stellar distances. Bessel's measurement of the parallax of 61 Cygni in 1838 was followed by other successes. By the 1870s astronomers, like the American Truman H. Safford (1836–1901), were arguing that "the stars' distances are inversely proportional *upon the whole* to their [the stars'] proper motions."[32] Many more stellar distances could be determined if proper motion data could be used directly in this way, because proper motion data were available in relative abundance: For each known trigonometric parallax there were scores of measured proper motions.

After 1875, the assumptions that stellar distances were inversely proportional to proper motions and that stellar motions were random increasingly influenced the direction of attack on the sidereal problem. This is not to suggest there was complete unanimity among astronomers. For example, the astronomers Maxwell Hall, Hugo Gyldén, and Eduard Schönfeld had independently proposed dynamic models of the stellar system in which random stellar motions were rejected and replaced by interstellar forces. Thus, in the words of Hall, "the Sun and stars are ... subject to the same law of Force, and revolve in immense orbits round the same centre."[33] But most astronomers, who were non-dynamicists like Argelander, Airy, and Safford, accepted the hypothesis of random motions applied to large groups of stars.[34]

Though the dynamic models were perhaps intellectually more appealing, particularly to theorists, the sheer immensity of the

---

[32] T. H. Safford, "On the solar motion in space and the stellar distances," *American Academy of Arts and Sciences, Proceedings*, 11 (1876), 55 (52–61) (second paper).

[33] M. Hall, "The sidereal system," *Royal Astronomical Society, Memoirs*, 43 (1876), 157 (157–97).

[34] T. H. Safford, "On certain groups of stars with common proper motions," *Royal Astronomical Society, Monthly Notices*, 38 (1878), 295 (295–7). It was generally admitted, however, that in some cases smaller groups of stars did exhibit preferential motions, a tendency at the time called "star-drifting."

sidereal problem prevented dynamicists from providing a satisfying conceptual basis needed to understand the arrangement of all the stars in space. In contrast, nondynamicists increasingly found empirical evidence correlating stellar motions with parallaxes. The publication by the German astronomer G. F. Auwers in 1888 of his reduction of Bradley's star catalogue, which listed 3,200 reliable proper motions, further stimulated these efforts. Nevertheless, prior to Kapteyn's studies of stellar motions during the 1890s, no one had succeeded convincingly in relating a distance measure based on a few thousand proper motions to the demands of the large survey catalogues, such as the *Bonner Durchmusterung*, containing hundreds of thousands of stars. Although it might well be complex, a precise relationship between proper motions, parallaxes, and apparent magnitudes, even if only statistically based, might be extremely useful. Such a relationship eventually provided the only reasonable alternative to Herschel's original brightness-nearness hypothesis. By century's end, these various issues led to Kapteyn's formula for stellar distances, which, in addition to the density and luminosity relationships and the "fundamental equation of stellar statistics," is the most significant result of the early developments in statistical astronomy.

Although Kapteyn generally preferred to use the distances directly derived from measured (trigonometric) parallaxes, because there were only a few hundred accurate parallax measurements by the turn of the century, he recognized that the solution of the sidereal problem, even with improved photographic techniques, demanded a much broader base than that allowed simply by the Earth's orbit. His emphasis on stellar motions was therefore motivated by both practical and theoretical considerations. From a practical point of view, by correlating proper motions with stellar parallaxes, the base of parallaxes could be extended on the ever-increasing baseline of the Sun's motion through space. From a theoretical point of view, for an understanding of the structure of the stellar system knowledge of only the mean distances of groups of stars, rather than the actual distances of individual stars, was necessary.

It appears that no matter how proper motions as a measure of distance were analyzed, however, either an observational or a conceptual limitation prevented a full generalization of the relationship. It was clear that stellar brightness as a measure of distance was invalid. But a reliable measure of proper motion (actually parallactic motion) appeared equally problematic. In 1892, Kapteyn and the Irish astronomer William H. S. Monck (1839–1915) independently discovered a direct correlation between proper motion and spectral type.[35] Also about the same time, astronomers discovered that using a broad range of proper motions would lead to spurious conclusions. Some argued, therefore, that a narrow range of motions would be a better approach.[36] Thus for groups of stars with proper motions between relatively narrow limits, the mean parallactic motion became a measure of the average distance (mean-parallax).

But Monck, Kapteyn, and others quickly pointed out, that a relationship between stellar motions and distances based on narrow limits was also not entirely valid.[37] Consider the following argument: Suppose two stars have an equal proper motion, say three radii of the Earth's orbit (three astronomical units) in a year, but in opposite directions. Because the Earth covers a distance of four radii per year,[38] it follows that the star moving in the same direction as the Earth will have a relative proper motion of only one radii, and will, therefore, not appear in our list as a star of large proper motion. On the other hand, the star moving with equal speed in the opposite direction will have a motion of seven radii per year, and will, therefore, be included among stars of considerable proper motion. Thus, a bias occurs in consequence

---

[35] J. C. Kapteyn, "Over de verdeeling van de sterren in de ruimte," *Verslagen en Mededeelingen der Koninklijke Akademie van Wetenschappen te Amsterdam Wis- en natuurkunde Afdeeling*, 9 (1892), 418–21; and W. H. S. Monck, "The Sun's Motion in Space," *Astronomical Society of the Pacific, Publications*, 4 (1892), 75–7.

[36] See J. G. Porter, "The Proper Motion of Certain Stars as a Criterion of their Distance," *Popular Astronomy*, 6 (1898), 549–53; J. C. Kapteyn, "On the Mean Parallax of Stars of Determined Proper Motion and Magnitude," *Groningen Publications*, no. 8 (1901), 1 (1–31).

[37] W. H. S. Monck, "The Sun's Motion in Space," *Astronomical Society of the Pacific, Publications*, 7 (1895), 33 (33–8); also see F. W. Ristenpart, "Untersuchungen über die Constante der Präcession und die Bewegung der Sonne in Fixsternsysteme," *Veröffentlichungen der Grossherzoglichen Sternwarte zu Karlsruhe*, 4 (1892), 287.

[38] Because the accepted value of the solar motion, ca. 1895, was about 20 km/ sec, the solar system covered a distance of about four astronomical units in a year.

of which we include many stars having a motion away from the solar apex, whereas the corresponding ones, necessary to cancel that motion, will be left out of the count. The parallactic motion will then, on the average, be too large in the case of the stars of large apparent proper motion.[39] For this and other reasons, Kapteyn began a systematic study of the relevant factors, and discovered around 1900 a relationship that formed the basis of nearly all of his subsequent work.

STATISTICAL COSMOLOGY: SEELIGER AND KAPTEYN

Throughout the nineteenth century, most stellar astronomers worked within the research tradition originally outlined by William Herschel. Although astronomers eventually recognized the limitations in these assumptions, for lack of alternatives they continued to emphasize the geometrical progression of stellar magnitudes. Because the empirical data and the theoretical models were never entirely in agreement, however, most astronomers modified their premises or added new ad hoc assumptions. Perhaps the most significant development of the period was the accurate cataloguing of large numbers of stars. This empirical data provided the raw substance for virtually all developments after mid-century.

Many astronomers also recognized that the sidereal system was far more complex than these assumptions allowed. For instance, debates concerning the "one island universe" and the nature of the Milky Way occupied a central position in nineteenth-century astronomical thinking. Dilemmas suggested by an infinitely extended universe containing an infinite number of stars challenged astronomers' conceptions of the sidereal system. Toward the end of the century, the problem of the gravitational and optical paradox of an infinite universe received considerable attention.[40]

---

[39] S. Newcomb, *The Stars* (New York: G. P. Putnam's Sons, 1901), p. 298.
[40] Newton was the first of many to appreciate the problems in these cosmological questions; see Hoskin, "Newton, Providence and the Universe of Stars," pp. 77–101. Those astronomers, among many others, who considered these difficulties in the nineteenth century include P. S. Laplace, W. Olbers, J. Herschel, Struve, and Seeliger. For relevant details see S. L. Jaki, *The Paradox of Olbers' Paradox: A Case History of Scientific Thought* (New York: Herder and Herder, 1969), pp. 144–98.

Nineteenth-century developments in this research tradition, however inadequately conceived for the task at hand, provided for an essential formative period that preceded developments in statistical cosmology after 1890. Though the nineteenth century ultimately added little of a substantive nature to this emerging science, it did provide a thorough discussion of the many problems and limitations of a highly suggestive – though relatively sterile – program. First, the assumptions of stellar distributions were thoroughly explored: These included, on the one hand, assumptions dealing with equal brightness and uniform distributions, and, on the other hand, those concerned with random stellar motions. Second, the massive star catalogues were produced with data for hundreds of thousands of stars. And third, various modeling schemes were devised to organize the stellar data in conformity with the basic assumptions. The latter development, particularly as a result of the work of Struve, Littrow, and Schiaparelli, brought a new level of sophistication to understanding how stars were thought to be distributed in space. Unfortunately, a lack of reliable methods for understanding the third spatial dimension prevented astronomers from solving the sidereal problem.

During the last years of the century, two innovative research programs emerged that finally overcame the limitations of these traditional approaches. In Germany Hugo von Seeliger developed a mathematical approach to the distribution of the stars, which was initially based on some of Schiaparelli's stellar work, that resolved the inadequacies of all earlier approaches to understanding stellar distributions. At the same time, in Holland the astronomer J. C. Kapteyn, using some sophisticated statistical ideas, devised methods based on data derived largely from stellar motions, as well as on the more familiar data available in stellar catalogues, to resolve the same dilemmas. As a result of Seeliger's and Kapteyn's programs, studies in the distributions and motions of stars, two interrelated but previously distinct research fields, merged into one coherent research tradition that provided the basis for virtually all studies in both statistical astronomy and statistical cosmology for the next four decades.

Although astronomical data of the most general kind were used in their attack on the sidereal problem, data dealing with the motions of stars, primarily proper motions, played a distinctly subsidiary role in the development of these statistical cosmologies. Following the gradual acceptance of the so-called big galaxy during the 1920s, however, stellar motions increasingly played a significant role in the emergence of the "new astronomy." This was particularly true with the discovery of "differential galactic rotation" in 1927, which, in effect, was a solution to the "two-stream" phenomenon Kapteyn discovered in 1902. The concept of galactic rotation was framed by Kapteyn's discovery, though it was heavily based on the data uncovered by statistical astronomers in the early years of the twentieth century, principally the work of the Swedish astronomers C. V. L. Charlier and his student G. B. Strömberg. Because our story deals with the nature of the Milky Way Galaxy a few years prior to the emergence of this newer cosmology, we will not focus on the significance of stellar motion data except insofar as it relates directly to the statistical cosmologies of Seeliger and Kapteyn.

Seeliger and Kapteyn were themselves not alone in developing these ideas. Both mentored academic and research astronomers and both developed well-defined centers of research: Significant empirical and conceptual problems had been clearly established; methodological approaches had been developed; dedicated research teams had been assembled; and formal means for distributing results had emerged. In the process, both Seeliger and Kapteyn defined characteristics that deeply reflected a unique style within their respective research communities. Though not formally connected elsewhere, a number of astronomers, particularly in Sweden, England, and the United States, also worked on problems most clearly conceptualized within this research tradition. In conclusion, statistical cosmology, as developed principally by Seeliger and Kapteyn, became a significant research endeavor, attracting relatively large numbers of astronomers who came to see it as a field with enormous possibilities for understanding the architecture of the Milky Way system.

Part II

STATISTICAL COSMOLOGY, 1890–1924

3

SEELIGER AND STELLAR DENSITY

In its most general form, the problem that Seeliger, Kapteyn, and others attempted to solve was to find how the stars are distributed through space according to absolute magnitude, spectral type, and distance from the solar reference. Although this general problem spawned a significant research program in astronomy with numerous research problems, Seeliger, Kapteyn, and their astronomical colleagues focused on different aspects of the problem that depended on a variety of astronomical, methodological, and institutional factors. These concerns will be dealt with in Chapters 3 and 4, Chapters 5 and 6, and Chapter 7 respectively.

Throughout his long and distinguished career, Seeliger addressed diverse scientific problems and carried on an active correspondence with numerous astronomers and scientists.[1] He explored many technical issues including topics as diverse as spiral nebulae, stellar evolution, Saturn's rings, universal gravitation, and, of course, statistical astronomy and cosmology. Among these, Seeliger focused on understanding one crucial problem, namely, the density relationship – expressing the absolute numbers of stars per unit volume of space – which he used to explore his central cosmological concerns. Even so, the major thrust of his work may be divided into two periods, each characterized by a different approach to basic questions. With the publication in 1884 of a study on the distribution of the stars in space, Seeliger began his investigations of stellar distributions that would eventually

---

[1] Seeliger corresponded on a wide range of scientific problems with a large number of astronomers worldwide, including David Hilbert, Ernst Mach, Arnold Sommerfeld, and Willie Wien. See, for example, J. Thiele, "Briefe Hugo v. Seeligers an Ernst Mach," *Philosophia Naturalis: Archiv für Naturphilosophie und die philosophischen Grenzgebiete der exakten Wissenschaften und Wissenschaftsgeschichte,* 17(3) (1979), 391–9, in which Seeliger addresses issues raised by Ludwig Boltzmann, Max Planck, and Mach himself.

consume his entire career. For the next decade and a half, he followed the general approach most others had hitherto advocated,[2] and formulated the sidereal problem in terms not unlike those expressed by Littrow, Gould, Schiaparelli, and others using the traditional methods that had been introduced by William Herschel and F. G. W. Struve. In other words, by examining star catalogues he attempted to determine the progression of the numbers of stars as a function of magnitude under the supposition that brighter stars are roughly closer, whereas fainter ones are farther away. Seeliger, and most others at the time, recognized the inherent problems built into this kind of analysis. But without additional data characterizing the distance to the stars, the sidereal problem remained virtually intractable. Then sometime around the turn of the century Seeliger radically changed his strategy and introduced some exceptionally powerful mathematical ideas that forever changed the landscape of this field of astronomical research. Briefly, these ideas were based on the (then) highly obtuse mathematics of integral equations. Such mathematics allowed Seeliger, and later Karl Schwarzschild and Kapteyn himself, to reformulate the problem of the distribution of stars into a coherent theoretical framework.

Seeliger was fully equipped to deal with this far more powerful mathematical methodology. The most important scientific influences on Seeliger were Carl Neumann, the outstanding mathematical physicist at Leipzig, and Carl Friedrich Gauss, the brilliant nineteenth-century mathematician.[3] In addition to perusing Neumann and Gauss, Seeliger studied the recognized classics in physical science – Jacobi, Lagrange, and Laplace – throughout his graduate career. At Leipzig, where Seeliger, received his doctorate in 1872, the astronomers E. H. Bruhns and P. A. Hansen, as well as Friedrich Zöllner, the first German professor of astrophysics

[2] H. Seeliger, "Die Vertheilung der Sterne auf der nordlichen Halbkugel nach der Bonner Durchmusterung," *Sitzungsberichte der Mathematisch–Physikalischen Klasse der K. Bayerischen Akademie der Wissenschaften zu München* (hereafter *München Ak. Sber.*), 14 (1884), 521–48.

[3] For background on Carl Neumann, see Christa Jungnickel and Russell McCormmach, *Intellectual Mastery of Nature: Theoretical Physics from Ohm to Einstein*, 2 vols. (Chicago: University of Chicago Press, 1986), vol. 1, pp. 181–5; on Gauss see Morris Kline, *Mathematical Thought from Ancient to Modern Times* (New York: Oxford University Press, 1972), pp. 870–1.

Figure 7   Hugo von Seeliger, ca. 1910. (Courtesy of Yerkes Observatory)

and the founder of modern photometry, significantly influenced his astronomical training.

Although Seeliger and Kapteyn were almost identical contemporaries and although they partially relied on the same data supplied by others, they worked almost entirely independently, though occasionally used one another's results. For example, Kapteyn used Schwarzschild's analyses, which in turn were based on Seeliger's work, as early as 1914, but only in his 1920 cosmology did Kapteyn use any of Seeliger's results directly and in a central way. Seeliger, who used some of Kapteyn's data as early as 1911, did not begin to respond to Kapteyn's work until about 1920. Given the close identity of many of their problems, their (non)relationship is indeed surprising. Briefly, the explanation lies in the fact that they advocated and used radically different approaches to the same problems. Because of their fundamentally different approaches, it is historically more accurate – and easier – to treat Seeliger and Kapteyn separately, virtually as though the other did not exist. In Chapter 6, which deals with statistical cosmology circa 1920, we will be in a position to explore questions of common interest among not only Seeliger and Kapteyn, but also the entire community of statistical astronomers.

THE EARLY YEARS

Because the *Bonner Durchmusterung* is restricted to the stars of the northern hemisphere and deals mostly with the telescopic stars brighter than magnitude 9.5, its publication did not preclude the execution of additional surveys. About the time that Littrow and Gould had published their investigations utilizing the *Bonner* catalogue, several other studies appeared based upon different stellar surveys. These other surveys were even less complete, however, and studies based upon them were correspondingly more restrictive. For instance, one study dealt with all stars brighter than magnitude 11.5 – nearly two full magnitudes fainter than the stars in the *Bonner Durchmusterung* – but it was limited to a zone only 6 degrees wide bordering the celestial equator. Another study considered the entire sky, but was restricted to the naked-eye stars. A third study dealt with the stars within 100 degrees of the south celestial pole, but down to stars of only magnitude 7.0.[4]

Questions concerning the density of stars and the spread of stellar brightness came no closer to a clear understanding than before. Strictly within a "statistical" approach to the problem of the distribution of stars, there appeared an increasingly wide divergence of opinion concerning the validity of Herschel's fundamental assumptions. As we have seen, some supported the conclusions represented by Littrow and Gould.[5] Others believed there was sufficient evidence "to finally dispose [sic] of the fundamental assumption that the stars are equally scattered in space."[6] By the mid-1880s, there was certainly no unanimity, one way or the other, concerning the ideas perhaps most clearly expressed by Littrow and Gould, even among those studies dealing exclusively with the counts of large numbers of stars.

Using additional stellar data besides that represented by the counts of stars (as a function of magnitude), it had become clear that, in fact, there were serious anomalies that Herschel's suppositions could not explain: (1) Stellar parallax as a measure of distance failed to confirm the proposition that "brightness implies nearness," because relatively close-by stars were, in some cases, fainter than more distant stars; (2) the "equal brightness" hypothesis was negated by considerations of binary star systems

[4] These three studies are, respectively, G. Celoria, "Sopra Alcuni Scandagli del Cielo eseguiti All'Osservatoire Reale di Milano, e sulla Distribuzione Generale delle Stelle Nell Spazio," *Publicazione del Reale Osservatoire di Brera in Milano*, no. 13 (1878), 3–47; J. C. Houzeau, "Uranometrie Generale, avec une Etude sur la Distribution des Etoiles visibles a l'Oeil Nu," *Annales de l'Observatoire de Bruxelles*, 1 (1878); and B. A. Gould, *Uranometria Argentina: Brightness and Position of every Fixed Star, down to the Seventh Magnitude within one Hundred Degrees of the South Pole* (Buenos Aires: Paul Emile Coni, 1879), chap. VIII, 348–83.

[5] Gould confirmed his earlier views (1874) in his 1879 study; see Gould, *Uranometria Argentina*, p. 369.

[6] E. S. Holden, "Statistics of Stellar Distribution derived from Star-Gauges and from the Celestial Charts of Peters, Watson, Chacornac, and Palisa," *The Observatory*, 7 (1884), 256 (249–56); also see W. H. S. Monck, "The Distribution of the Stars," *The Observatory*, 7 (1884), 296–7, and W. H. S. Monck, "A Note on the Distribution of the Stars," *Sidereal Messenger*, 7 (1888), 20–5.

in which the component stars differed greatly in terms of their intrinsic brightnesses; and (3) the hypothesis of "equal stellar distribution" was challenged by the very existence of stellar clusters. There were a few astronomers who attempted to resolve some of these anomalous features by means largely foreign to statistical considerations using stellar counts.[7] These attempts received little support, however, because in general they did not have as their central concern an explanation of the sidereal system, which, in the opinion of most astronomers at the time, could be achieved only by a statistical understanding of large numbers of stars. Several attempts, including those most successfully initiated separately by Seeliger and Kapteyn, were made over the following two decades to explain these anomalies largely by statistical analyses of the counts of stars as a function of their apparent magnitude.

Despite Littrow's early efforts, Seeliger was the first to exploit fully the *Bonner Durchmusterung*. Above all else, he recognized the singular importance of reliable data: Accurate, comprehensive counts of stars were absolutely crucial in any study of the distributions of the telescopic stars. As Littrow had done earlier, he immediately turned to the most comprehensive source for his empirical data, the *Bonner Durchmusterung*. The lack of a comparable survey of telescopic stars in the southern hemisphere was unfortunate, because a complete understanding of the problem would require more than the incomplete data then available.[8] It was well known within the astronomical community, however, that since 1877 Eduard Schönfeld (1828–1921), director of the Bonn Observatory, had been engaged in the process of filling this

---

[7] See E. C. Pickering, "Dimensions of the Fixed Stars, with Special Reference to Binaries and Variables of the Algol Type," *Proceedings of the American Academy of Arts and Sciences*, 16 (1881), 1–37; and E. S. Holden, "On some of the Consequences of the Hypothesis, Recently Proposed, that the Intrinsic Brilliancy of the Fixed Stars is the Same for Each Star," *Proceedings of the American Association for the Advancement of Science*, 29 (1880), 137–51. Also during the 1870s T. H. Safford attempted to replace stellar brightness with proper motion as a measure of distance; see T. H. Safford, "On the Solar Motion in Space," *Proceedings of the American Academy of Arts and Sciences*, 10 (1875), 82–90; 11 (1876), 52–61; and 11 (1876), 210–17.

[8] H. Seeliger, "Ueber die Vollständigkeit wiederholt ausgefuhrter astronomischer Durchmusterungsarbeiten," *Astronomische Nachrichten*, 104 (1883), 225–46.

empirical lacuna by extending the *Bonner* survey to 23 degrees south declination.[9] Seeliger would have preferred a complete stellar survey for his research, but he knew that goal was not realistic; the task involved in obtaining such a survey was measured in terms of years and even decades. Indeed, the telescopic survey was not completed until 1930, though a photographic survey was obtained by 1899.[10]

Rather than attempt to verify certain premises as Littrow, Gould, and others had done, Seeliger emphasized a methodological approach that was designed to infer empirical regularities suggested by the data. His approach was not strictly inductive, though, because he was aware of certain regularities in the stellar phenomena suggested in the studies of other astronomers. For instance, it had become evident, since the time of William Herschel's studies, that the Milky Way was not a local phenomenon; rather, the stellar system appeared to be largely symmetrical about the Milky Way and somehow related to it.

Seeliger's early research on stellar distributions focused on the importance of the Milky Way for the *entire* sidereal system. Star positions in the *Bonner* catalogue were recorded in terms of the equatorial coordinates, right ascension and declination, and therefore star positions were inclined by 62.5 degrees to the galactic plane of the Milky Way.[11] Because Seeliger was interested in the counts of stars with respect to the Milky Way, he found it necessary to transform *each* of the *Bonner* stars, over 324,000, from the equatorial to the galactic coordinate system.

It was evident to Seeliger that the effort for such an ambitious project would certainly be "tiresome and wearisome." Littrow, who had been the only astronomer before Seeliger to count the *Bonner* stars, had simplified the arduous task of counting simply

[9] E. Schönfeld, "Berichte uber die Thätigkeit der Sternwarten in Bonn," *Vierteljahrsschrift der Astronomische Gesellschaft,* 12 (1877), 41–55; 13 (1878), 119–25; 14 (1879), 118–22; 15 (1880), 93–6; 16 (1881), 200–3; 17 (1882), 188–90; 18 (1883), 81–3.

[10] *Córdoba Durchmusterung* (1930), and D. Gill and J. C. Kapteyn, *Cape Photographic Durchmusterung* (1896–1990).

[11] Seeliger used the following equatorial coordinates for the celestial north pole: R. A.: 12h, 49m; D: +27.5 degrees; see Seeliger, "Die Vertheilung der Sterne auf der nordlichen Halbkugel nach der Bonner Durchmusterung," p. 541.

by ignoring altogether the right ascension coordinate position of the stars. By doing so, however, Littrow failed to consider the longitudinal coordinate in his earlier work, a fact that Seeliger deeply regretted:

> So worthless is this work for the statistics of the stars, that only very little can be said concerning the distribution of stars because the right ascension was not considered. It is therefore regrettable that Littrow considered his work to be important. It would have been much more interesting, if it had been partitioned along the right ascension, because this could have been accomplished with proportionately little trouble. Now the work must be executed anew and Littrow's counts can only be used as a control.[12]

Because Seeliger was interested in the Milky Way system as a whole, his program demanded that the right ascension coordinate system be transformed into the galactic coordinate system.

Seeliger developed an ingenious scheme that allowed him to avoid the tedious and time-consuming work that would have been required to perform coordinate transformations for each of the *Bonner* stars individually. Rather than calculate the coordinate transformation for each star (an impossible time-consuming task), Seeliger counted the number of *Bonner* stars as a function of magnitude class, declination class, and right ascension class – still an arduous task. Combining these results into small trapezoids, which divided the entire northern hemisphere bounded by known values of right ascension (4 minutes wide) and declination (1 degree high), Seeliger determined the number of stars of each magnitude class within each trapezoid. Starting at the galactic north pole, Seeliger divided the northern sky into eight successive zones parallel to the galactic equator (or Milky Way) such that each zone had an altitude of 20 degrees, except the eighth zone, which spanned the distance from 50 degrees south latitude down to the galactic south pole.[13] Thus the fifth zone, from 10 degrees

---

[12] Ibid., pp. 521–2. Unless otherwise indicated, all translations are mine.
[13] Seeliger recognized that J. C. Houzeau had divided the celestial sphere into the same zones six years earlier in a study on the distribution of the naked-eye stars; see Houzeau, "Uranometrie Generale," pp. 32–7.

north to 10 degrees south latitude, contained the plane of the Milky Way. Finally Seeliger intersected the northern hemisphere with the eight galactic zones and calculated the equatorial coordinates of the boundaries of each of these zones. Because each zone was obliquely inclined by 117.5 degrees (90 + 27.5 degrees in the positive direction) to the celestial equator, all of the small trapezoids, except for a few that straddled two zones, were contained entirely in one of the zones. Making the necessary adjustments for the few irregular trapezoids, Seeliger summed the counts of stars of all trapezoids within each zone, thereby determining the total number of stars in all of the eight zones as a function of magnitude class

In this new form the data immediately confirmed the supposition that the number of stars in each of the magnitude classes increases as one approaches the Milky Way from either of the galactic poles. This provided the first firm quantitative confirmation using the *Bonner* stars of this long-observed stellar phenomenon. Moreover, using an earlier study by Jean-Charles Houzeau, Seeliger found the same relationship to exist among the naked-eye stars with respect to the Milky Way. On the basis of these empirical conclusions, he conjectured that an identical relationship exists for all stars whose magnitude is greater (i.e., fainter stars) than the *Bonner* limiting magnitude of 9.5.

Seeliger extended his investigations into the southern hemisphere two years after the publication of his first treatise.[14] His new study was based on Schönfeld's survey, which had been published early in 1886 as the sequel to Argelander's *Bonner Durchmusterung*.[15] Schönfeld's catalogue contains all stars down to the tenth magnitude situated between 2 and 23 degrees south declination, thus including nearly the entire ecliptic. After obtaining a copy of the *Southern Durchmusterung* from Schönfeld, Seeliger completed an analysis of its contents that was similar to that found in his 1884 paper. He easily confirmed the existence of his earlier conclusions. In addition, he determined empirically, as far as could be inferred from the data of both *Durchmusterungen*, that on the

---

[14] H. Seeliger, "Ueber die Vertheilung der Sterne auf der sudlichen Halbkugel nach Schönfeld's Durchmusterung," *München Ak. Sber.*, 16 (1886), 220–51.

[15] E. Schönfeld, *Bonner Sternverzeichniss* (Bonn: Adolph Marcus Verlag, 1886).

average the distribution of stars is symmetrical about the Milky Way.

The conclusions that Seeliger had thus far proposed, however, were relatively insignificant. Important results that could add to the understanding of the stellar system required, above all, accurate determinations of stellar magnitudes, not only positional surveys of the stars:

> I believe that not much can be gained simply through surveys such as those which have been recommended in recent times. Based on these one can make only slightly interesting claims concerning the apparent distribution of the stars on the celestial sphere. Without dependable data on the brightness of stars, one cannot draw further conclusions.[16]

Thus the immediate pressing need was to determine the accuracy of the magnitude determinations in both Bonner catalogues, preferably from an independent source. Still, these two catalogues provided the most comprehensive surveys of the telescopic stars achieved by Seeliger's time. Seeliger recognized the importance of his transformations for future statistical investigations, and therefore in 1891 he published a complete summary catalogue of the star-counts found in the 1884 and 1886 papers, which formed the basis of the highly innovative work he undertook through much of his later career.[17]

STATISTICAL COSMOLOGY AND UNIVERSAL GRAVITATION

Of greater significance for the future direction of the subject, however, is the fact that during the late 1890s Seeliger began to reassess the traditional approach used in statistical studies of the distribution of the stars. There appear to be a number of reasons that influenced Seeliger to undertake a radical restructuring of

[16] Seeliger, "Ueber die Vertheilung der Sterne auf der sudlichen Halbkugel nach Schönfeld's Durchmusterung," p. 235.
[17] H. Seeliger, "Die Vertheilung der in beiden Bonner Durchmusterung enthalten Sterne am Himmel," *München Sternwarte Neue Annalen*, 2, Part 1891, Part C, 42 pp.

stellar statistics: (1) his mathematical propensity, (2) his relationship to the Swedish astronomer, Hugo Gyldén, and (3) his critique of Newtonian physics.

Seeliger understood thoroughly the importance of observational work to astronomy, having acquired his commitment to the observational side while he completed his astronomy doctorate under E. H. Bruhns at Leipzig, further refining his practical understanding of the field later as Argelander's assistant at the Bonn Observatory. Even so, he was heavily predisposed early on to the use of highly abstract mathematics, which he acquired from his scientific mentor and real inspiration, Carl Neumann, the outstanding mathematical physicist who was also at Leipzig. This combination of theoretical understanding and observational training with severe attention to minute detail partially explains, on the one hand, such "tiresome and wearisome" projects as counting the *Bonner* stars, and, on the other hand, building a theoretical synthesis in statistical cosmology. Although Seeliger shared his observational side with most astronomers, his connections with Neumann and his propensity toward the abstract present us with an astronomer who at least equally advocated a mathematical and theoretical foundation to the discipline, for example, such as what emerged in his studies of statistical cosmology beginning around the turn of the century.

In 1883 Seeliger was elected first secretary of the prestigious *Astronomische Gesellschaft,* where he served until 1896 when he became its president following the death of Hugo Gyldén, who had been its president since 1889. There is numerous personal correspondence between the two astronomers, as well as announcements of policy, operation, and so on over the signatures of both scientists in the society's proceedings, the *Vierteljahrsschrift.*[18] Although this is not in itself out of the ordinary, the fact that Seeliger's approach to the question of the distribution of the stars (beginning in 1898) followed Gyldén's general understanding (published in a paper 1872) in several important details is remarkable. First, both men had published in the field of stellar distributions basing their research on the *Bonner* data; second, in both cases their articles began with an extensive discussion of

[18]   H. v. Seeliger to H. Gyldén, 1884–96, passim (Stockholm).

Littrow's investigations; and third, and most significant, both astronomers – and only these two astronomers! – used the nascent theory of integral equations to analyze their problems (the general theory of integral equations was not placed on a sound foundation until 1896–7). Most important, Gyldén suggested that the luminosity function might be derived using a first-order integral equation. Although Gyldén never solved the general problem, and Seeliger (some twenty-six years later) proposed deriving luminosity and density functions using second-order integral equations, it is most curious that only these two astronomers adopted this particular approach – until, of course, Seeliger succeeded in making integral equations the foundations of the study of stellar distributions. Once Seeliger reduced the problem to integral equations, his solution was highly original and justifiably his.

During the 1890s a number of astronomers began to reconsider the formulation of Newton's law of gravitation. They began to believe that questions dealing with the validity of Newton's gravitation relationship on a cosmological scale hinge ultimately on the mass of the sidereal universe. Consequently, issues dealing with the distribution of the stars are central, because depending on the density gradient of stellar space, information on the mass of the universe is needed to legitimate Newton's law.[19] As Seeliger came to question the adequacy of Newton's gravitation law, he realized that a careful examination of the distribution and mass (luminosity) of stars were the crucial data needed to determine this fundamental condition of the cosmos. With thirty years of training and preparation behind him, Seeliger was in a position as one of few astronomer-cosmologists at the turn of the century to understand what was at stake: nothing less than an understanding of the cosmic glue holding together the very fabric of the Universe!

Specifically, Seeliger was concerned about the applicability of

[19] Seeliger discussed his ideas with many of the most capable physicists and mathematicians in Germany, including Arnold Sommerfeld, who was among the first to support Einstein's theory of relativity, thus rejecting Newtonian gravitational theory. H. Seeliger to Arnold Sommerfeld, 25 May 1902 (Deutsches Museum).

Newton's law on an *infinitely* extended stellar system. The notion of infinity, both mathematically and physically, had been almost universally rejected throughout the nineteenth century. Had it not been for the fact that Seeliger devoted considerable effort during the 1890s to refute the notion of an actual infinity, we might simply interpret his views as an expression of his time. His rejection of a completed infinity, however, must be seen as far more than casual. Not only did he devote several very technical scientific papers specifically to the issue between 1894 and 1896, but with the publication of his first cosmological conception of his stellar system in 1898, he also found a direct way to defend his views of a *finite* universe.

The nineteenth-century rejection of an actual infinity has its roots in the eighteenth century when mathematicians freely used infinitely large numbers as a result of a less rigorous understanding of the foundations of mathematics. It was left to the founders of modern analysis, particularly the French mathematician Augustin-Louis Cauchy, to divest the idea of infinity of its metaphysical ties and formally to reject the existence of an actual infinity. Gauss, who had been one of the most significant influences on Seeliger's intellectual development, wrote: "I protest against the use of an infinite quantity as an actual entity, this is never allowed in mathematics. The infinite is only a manner of speaking. . . ."[20] Gauss's views on infinity were widely known and permeate his mathematical works.[21] Still, during the second half of the century, although developments in set theory did not alter the status of the infinite, they did introduce a viable usage of the idea of infinity into mathematics. In 1851 the Bohemia mathematician and philosopher Bernhard Bolzano, in his *Paradoxes of the Infinite,* defended the existence of actual infinite sets. And by the 1870s, the German mathematician Georg Cantor was beginning to develop his powerful ideas about the theory of sets, which were predicated on the existence of an actual infinite. Cantor's ideas, however, were savagely attacked by his compatriot Leopold Kronecker, whereas Felix Kline, Henri Poincaré, and many others

[20] K. F. Gauss, *Carl Friedrich Gauss Werke* (Leipzig, 1900), vol. 8, p. 216.
[21] Kline, *Mathematical Thought from Ancient to Modern Times,* p. 993.

left mathematicians generally suspicious of Cantor's work well into the twentieth century.[22] Not until his death in 1914 did Cantor's ideas sanction the introduction of completed infinities into cosmological speculations.[23]

Seeliger's general understanding of current mathematical thinking certainly exposed him to the mathematical community's deep suspicion of the idea of an actual infinite. The mathematician's notion of actual infinity, however ubiquitous it may have been, is still not quite the same as that understood in physics and cosmology. Seeliger was primarily concerned with the *cosmological* implications of an infinite universe, and not with its mathematical formulation. Specifically, it was Newton's law of universal gravitation which, in Seeliger's mind, posed certain unacceptable paradoxes of an infinitely extended universe.[24]

Although Seeliger's understanding of the difficulties implicit in the idea of an actual infinity was encouraged by this nineteenth-century mathematical tradition, his direct source for his solution came primarily from physics. Implied in Newton's general law of universal gravitation, argued Seeliger, was the problem that in an infinitely extended universe uniformly filled with an infinite quantity of mass, one is demonstrably led to the unacceptable conclusion that an infinite gravitation force exists everywhere in the Universe.[24] Seeliger demonstrated mathematically that in such a system every body would be attracted by an infinite pull – infinite not in the metaphorical sense of something very large, but in the mathematical sense of being larger than any number we could imagine:

> [I]nfinitely great accelerations must occur in the universe, and this would be true with any conceivable mode of mass-distribution whatever. There would therefore result motions which, starting from finite velocities, would lead within a finite time, to infinite velocities. This

[22] Ibid., pp. 994–1003.
[23] J. D. North, *The Measure of the Universe: A History of Modern Cosmology* (Oxford: Clarendon Press, 1965), pp. 377–8.
[24] For a history of this problem, see G. J. Whitrow, "From the Problem of Fall to the Problem of Collapse: Three Hundred Years of Gravitational Theory," *Philosophical Journal: Transactions of the Royal Philosophical Society of Glasgow*, 14 (1977), 67–84.

conclusion contains within itself either an absurdity, or a direct contravention of the theory of mechanics.[25]

Called the "gravitational paradox," physical reality as we know it simply could not exist in that kind of universe. Not only was such a universe "not representable," but Seeliger realized that it was also not "entirely independent of metaphysical speculation."[26]

In order to maintain the gravitational potential and avoid the catastrophic collapse of the Universe, Seeliger modified Newton's inverse square law by introducing an exponential correction factor $- e^{-mr}$, where m is a quantity sufficiently small so that the transformed equation is insignificant except at very large distances.[27] This very correction, however, had already been proposed in 1825 by the French mathematical-cosmologist Laplace in his *Mecanique celeste*.[28] Seeliger had been raised on the classics of celestial mechanics and with Argelander's assistance in Bonn he perused Laplace's great work between observations during his nightly assignments.[29] It is therefore unlikely that he would have failed to have become thoroughly familiar with Laplace's solution, particularly as Laplace discussed it in a chapter entitled "Sur le loi de l'attraction universelle." Furthermore, in 1874 Carl Neumann, who had been Seeliger's most significant scientific influence, had already discussed the problem of Newton's law.[30] When Seeliger first discussed these problems during the 1890s, he even noted that Neumann, who was faced with the same concerns of the inverse square law of attraction relating to electrostatics, had proposed a solution identical to the one suggested later by

[25] H. Seeliger, "On Newton's Law of Gravitation," *Popular Astronomy*, 5 (1897/98), 546 (544–51).

[26] H. Seeliger, "Ueber das Newton'sche Gravitationsgesetz," *Astronomische Nachrichten*, 137 (1895), 132–3 (129–36). Also, see K. Laves, "On Some Modern Attempts to Replace Newton's Law of Attraction by Other Laws," *Popular Astronomy*, 5 (1898), 513–18.

[27] Seeliger, "Ueber das Newton'sche Gravitationsgesetz," (1894), pp. 129–36; and H. Seeliger, "Ueber das Newton'sche Gravitationsgesetz," *München Ak. Sber.*, 26 (1896), 373–400.

[28] P. S. Laplace, *Mecanique celeste* (1825), vol. 5, pp. 445–52.

[29] Alexander Wilkins, *Hugo von Seeliger's Wissenschaftliches Werk* (München: Verlag der Bayerische Akademie der Wissenschaften, 1927), p. 5.

[30] Carl Neumann, "Ueber die Helmholtz'sche Constant k," *Berichte über die Verhandlungen der Mathematisch-Physischen Classe der K. Sachsischen Gesellschaft der Wissenschaft Leipzig*, 26 (1874), 132–52.

Seeliger.[31] Because Seeliger studied with and thoroughly mastered the works of Neumann, as well as those of Laplace and Gauss, there can be little doubt that these sources combined to provide the basis for his understanding of the Newtonian gravitational and mass-density paradoxes.[32]

Seeliger had refused to ignore these difficulties altogether, and therefore he sought nothing less than a reformulation of Newton's theory. He was not willing, however, to expunge Newton's gravitational law entirely. As the nineteenth-century polymath John T. Merz has observed, "so great has the coincidence of calculation with observation turned out to be in all problems of physical astronomy, that no astronomer at the end of the century doubts that the gravitation formula alone will suffice to explain all anomalies which still exist in great number in the movement of cosmic bodies."[33] Prior to Seeliger's investigations, the history of efforts to address the gravitational and optical paradoxes before relativistic solutions in the early years of the twentieth century had all assumed that the Universe must be extended infinitely. That dogma, as one specialist has recently argued, has its origins in the "philosophical and quasi-theological" view to minimize the contingency of the Universe.[34]

STELLAR COSMOLOGY AND STATISTICAL ASTRONOMY

For Seeliger the issue remained unresolved: The Universe is either infinite, in which case Newton's law needed modification, or the Universe is finite, in which case the understanding of the stellar universe needed drastic revision. What appears to have motivated

[31] Carl Neumann, *Allegmeine Untersuchungen über das Newton'sche Princip der Fernwirkung mit besonderer Rucksicht auf die elektrischen Wirkungen* (Leipzig, 1896), pp. 1–3; also see Carl Neumann, *Untersuchungen über das Logarithmische und Newton'sche Potential* (Leipzig, 1877), pp. 69–78.

[32] See North, *The Measure of the Universe*, pp. 16–18, and S. L. Jaki, *The Paradox of Olbers' Paradox: A Case History of Scientific Thought* (New York: Herder and Herder, 1969), pp. 192–3.

[33] John T. Merz, *A History of European Thought in the Nineteenth Century* (New York: Dover Publ., 1965), vol. 1, part 1, p. 375.

[34] Stanley L. Jaki, "Das Gravitations-Paradoxon des unendlichen Universums," *Sudhoffs Archiv für Geschichte der Medizin und der Naturwissenschaften*, 63 (1979), 105–22.

Seeliger, therefore, was fundamentally nothing less than the construction of a cosmology that could resolve this issue in Newtonian celestial physics and in the process provide a detailed view of the actual workings of the cosmos. By the late 1890s, Seeliger was in a position to engage his self-appointed task: (1) He had thoroughly acquainted himself with the technical literature dealing with the distribution of the stars; (2) he was mathematically sophisticated, having acquired the necessary mathematical tools, and, most important, (3) believing he understood the inadequacies of the Newtonian view, he felt that he could explore the nature of the cosmos by a careful mathematical examination of the sidereal problem.

Beginning in 1898 Seeliger laid the basis for his innovative work in stellar theory and cosmology that would eventually consume his most important and productive years as a statistical astronomer and observational cosmologist by reducing the two *Bonner* catalogues as well as his own summary catalogues of the counts of stars. As they stood, however, these catalogues were still not suitable for his purposes. A decade earlier Seeliger had already noted that "one can make only slightly interesting claims concerning the apparent distributions of the stars. . . . Without dependable data on the brightness of stars, one cannot draw further conclusions."[35] Notwithstanding the many advances in observational astronomy, and given the relatively inaccurate though crucial data of the *Bonner* surveys, Seeliger pressed for a reduction of the data that would yield accurate and consistent magnitudes. The first significant attack on this problem had been outlined by Edward Pickering in his 1890 "Photometric Revision of the *Bonner Durchmusterung*," which unfortunately contained results for only a relatively small set of accurately known apparent magnitudes. Because Seeliger's research program required a *complete* set of corrected magnitudes for all the *Bonner* stars, he commenced with the necessary reductions and published them in 1898.[36] With the publication of these data, astronomers obtained accurate stellar magnitudes of the fainter stars needed for reliable

[35] Seeliger, "Ueber die Vertheilung der Sterne auf der sudlichen Halbkugel nach Schönfeld's Durchmusterung," p. 235.
[36] H. Seeliger, "Ueber die Grossenklassen der telescopischen Sterne der Bonner Durchmusterungen," *München Ak. Sber.,* 28 (1898), 147–80.

analyses of the distributions of stars.[37] Commenting on this and subsequent studies, Arthur S. Eddington wrote in his 1914 classic on the structure of the Universe: "The value of any compilation of star-counts will depend mainly on the accuracy with which the magnitudes, to which they refer, have been determined."[38] In 1898 Seeliger published, based on these reductions, his "Considerations on the Distribution of the Fixed Stars in Space," the first of five major treatises on cosmology.[39] His second study of 1898 initiated the modern era in statistical cosmology, and in it Seeliger outlined mathematico-statistical techniques that he – and later others – used so effectively in the construction of various statistical cosmologies.

The study is conceptually divided into two parts: The first deals with the exploration of a variety of empirical regularities, whereas the second provides a mathematical theory to explain those regularities. Using the corrected magnitudes of the *Bonner* stars, the first part of his study is based on the "star-ratio" (see Appendix I), or what is also called the "density-ratio," or using Eddington's phrase, the "fundamental theorem," relating to the statistics of stellar counts: Namely, that in a stellar system of unlimited extent in which the stars are distributed *uniformly,* the ratio of the number of stars of a given apparent magnitude, say 9.0, to the number of stars one magnitude brighter, 8.0, is 3.981.[40] This theoretical value, first exploited extensively as we have seen by Schiaparelli, assumes negligible interstellar absorption and a stellar absolute brightness independent of the distance from the Sun, problems that Seeliger would repeatedly address in subsequent studies.

Seeliger suggested several critical tests of the fundamental theorem's assumptions that could be examined rigorously with the

---

[37] See H. Kobold, "Literische Anziehen – Seeliger's 'Ueber die Grossenklassen der telescopischen Sterne der Bonner Durchmusterung'," *Vierteljahrsschrift der Astronomische Gesellschaft,* 34 (1899), 192–8.

[38] A. S. Eddington, *Stellar Movements and the Structure of the Universe* (London: Macmillan & Co., 1914), p. 185.

[39] H. Seeliger, "Betrachtungen über die räumliche Verteilung der Fixsterne," *Abhandlungen der Mathematisch-Physikalischer Klasse der K. Bayerischen Akademie der Wissenschaften zu München (hereafter München Ak. Abh.)* 19 (1898), 565–629.

[40] Eddington, *Stellar Movements,* pp. 184–5.

empirical data he now possessed. Because any scientific cosmology worth its weight should be able to say something about the distribution of the stars, Seeliger argued that if the empirical value for the star-ratio fell below its theoretical value, then the spatial penetration in the line of sight should be so great that a thinning out in the stellar density of the distribution of stars should be easily detected. Seeliger framed the question in terms of a relationship between a diminishing star-ratio and the direction of observation as measured by the galactic latitude. As such, Seeliger believed that the cosmology of the stellar system would reveal (1) the flattening of the stellar system toward the plane of the Milky Way, (2) the relative extent of the stellar system, and (3) the general distribution of stars as a function of magnitude and galactic latitude.

By exhaustively analyzing the statistical data of star-counts, stellar magnitudes, and galactic latitude with respect to the star-ratio, Seeliger discovered several empirical regularities. First, he found that "the counts of stars between 6.0 and 9.0 magnitude increase considerably slower with increasing magnitude than demanded by the [fundamental] theorem and its underlying assumptions."[41] Under the assumption of negligible interstellar light absorption, this result indicated that, on the average, the stars must be more sparsely distributed than had been theoretically expected. Second, Seeliger noticed that "the counts of stars increase all the more in proportion to magnitude the closer they approach the plane of the Milky Way."[42] Although Seeliger (1884–6) and Schiaparelli (1889) had already noticed the increase in the numbers of stars as the Milky Way is approached from either of the galactic poles, thus indicating that the Milky Way is not a local event, Seeliger's second result rigorously confirmed this phenomenon simultaneously for both increasing stellar magnitude *and* decreasing galactic latitude. Together these conclusions strongly indicated a thinning out in the density of the distribution of the stars in all directions of the sidereal system, though less so toward the plane of the Milky Way. Consequently, the stellar system appeared finite in extent and flattened toward the galactic poles.

[41] Seeliger, "Betrachtungen über die räumliche Verteilung der Fixsterne," (1898), p. 576.
[42] Ibid., p. 579.

Seeliger hastened to add that these first two conclusions were rigorously based only on the *Bonner* stars, those between 6.0 and 9.0 magnitude. Returning to the incomplete data from the catalogues of Giovanni Celoria and the Herschels,[43] Seeliger found strong evidence that his conclusions could still be qualitatively extended to the stars up to magnitude 11.5, whereas stars fainter yet to about magnitude 13.5 suggested a more rapid increase in the star-ratio, particularly toward the Milky Way. These observations constituted his third result that "the number of faint stars increases very slowly in regions far from the Milky Way and in an exceedingly much slower proportion [i.e., a higher ratio] than is the case with the brighter stars."[44] "These counts indicate ... with an absolute certainty," concluded Seeliger, "that the distribution of the stars essentially fainter than Celoria's stars [i.e., Herschel's] show an entirely different relationship than the brighter stars."[45]

Although they moved the discussion forward, his analyses were cautious because he understood fully the paucity of some of the data, particularly Celoria's and Herschels' data, and the fact that he was dealing with statistical averages. The conclusions, he wrote, "are true only for mean values that related to large areas. Single regions of smaller size often show quite different behavior; this appears especially to be the case toward the Milky Way."[46] Still, Seeliger's regularities received considerable attention. By 1906 the German astronomer Hermann Kobold had referred to them as empirical laws, and within a decade they had been received into the literature of statistical astronomy as fundamental relationships.[47]

---

[43] G. Celoria, "Sopra Alcuni Scandagli del Cielo eseguiti All'Osservatoire Reale di Milano, e sulla Distribuzione Generale delle Stelle Nell Spazio," pp. 3–47; E. S. Holden, "The Star-Gauges of Sir William Herschel, reduced to 1860," *Washburn Observatory Publications*, 2 (1884), 113–73, and J. Herschel, *Results of Astronomical Observations* (London: Smith, Elder & Co., 1847), pp. 375–9.

[44] Seeliger, "Betrachtungen über die raumliche Verteilung der Fixsterne," (1898), p. 593.

[45] Ibid.   [46] Ibid., p. 589.

[47] H. Kobold, *Der Bau des Fixsternsystem* (Braunschweig, Germany: Friedrich Vieweg und Sohn Verlag, 1906), see esp. pp. 163–5, and W. J. A. Schouten, *On the Determination of the Principal Laws of Statistical Astronomy* (Amsterdam: W. Kirchner, 1918), pp. 54–61. For an early discussion of

By the beginning years of the twentieth century, relatively little raw empirical data about the sidereal system was available other than star-counts (which included stellar positions and magnitudes), some proper motions, and the parallaxes of the brighter stars. To the astronomer interested in the structure of the Universe, this was an incredibly frustrating situation. As late as 1914 Eddington could write with confidence, albeit with some exaggeration: "Beyond the tenth magnitude there is an ever-increasing multitude of stars, which hold their secrets securely. We know nothing of their parallaxes, nothing of their spectra, nothing of their motions. There is only one thing we can do – count them."[48] The counting of stars first became theoretically useful with the ordering of the raw data using the fundamental theorem. With this in mind, Eddington recognized that "carefully compiled statistics of the number of stars down to definite limits of faintness can still yield information which is of value for our purposes."[49] In Seeliger's hands this proved particularly true.

Although Seeliger's initial results required various simplifying assumptions, in time he and others would modify and even reject them as additional empirical data became available. For example, Seeliger, Kapteyn, and a host of others recognized that the existence of interstellar light absorption could profoundly affect their understanding of the distribution and brightnesses of stars. In light of the fact that the existence of absorption was not confirmed until 1930, however, there would have been no scientific basis to accept one absorption function over another.[50] Invoking the principle of parsimony, Seeliger argued that given equally arbitrary assumptions the least problematic alternative – that of no absorption – was certainly the most desirable.

Not all assumptions, however, were so easily dismissed.

Seeliger's three conclusions, see H. Kobold, "Literische Anziehen – Seeliger's 'Betrachtungen über die räumliche Verteilung der Fixsterne', "*Vierteljahrsschrift der Astronomische Gesellschaft*, 34 (1899), 198–215, and H. Seeliger, "Zur Verteilung der Fixsterne am Himmel," *München Ak. Sber.*, 29 (1899), 363–413.

[48] Eddington, *Stellar Movements*, p. 184.   [49] Ibid.

[50] R. J. Trumpler, "Preliminary Results on the Distances, Dimensions and Space Distributions of Open Star Clusters," *Lick Observatory Bulletin*, 14 (1930), 154–88, and R. J. Trumpler, "Absorption of Light in the Galactic System," *Astronomical Society of the Pacific, Publications*, 42 (1930), 214–7.

Fundamentally, an understanding of the distribution of the stars in space requires knowledge of both the density and the luminosity laws. As William Herschel had known so well a century earlier, taken together these relationships form a vicious circle: Complete knowledge of one requires knowledge of the other. The legacy of nineteenth-century statistical studies testifies to the uncompromising nature of this dilemma. The traditional solution, of course, was to assume that the stars are equally bright. In Seeliger's view, this assumption constituted the essential weakness of all previous investigations of the distribution of the stars: "The inadmissibility of the hypothesis of equal illuminating power of all stars has already long been emphasized and indeed is above all the appropriate assumption to replace."[51]

As we mentioned earlier, except for only a relatively few stellar parallaxes and proper motions, the only data available in quantity for the analysis of stellar distributions were star-counts derived from the positions and apparent magnitudes of stars. As an observational cosmologist, Seeliger's primary goal was to understand the spatial distribution of the stars. Traditional assumptions, particularly the equal brightness hypothesis, were of no use. As a consequence he replaced this brightness assumption with an unspecified probability function, thus emphasizing that *all* stellar luminosities from the faintest to the brightest are possible within any given volume of space and that their luminosity distribution may be specified when the data becomes available (see Eq. 1, the probability luminosity function).

$$\int_0^H \Phi(i)di = 1 \tag{1}$$

This mathematical device allowed Seeliger to circumvent the density–luminosity dilemma. Although this probability function expresses the most general view that all luminosity values are possible, it did not provide the analytic form of the luminosity function itself. In order to move the discussion beyond the usual stalemate in statistical cosmology, Seeliger temporarily replaced the probability function with a fixed analytic form forcing the same distribution of luminosities to exist everywhere in space. In

[51] Seeliger, "Betrachtungen über die räumliche Verteilung der Fixsterne," (1898), p. 594.

so doing, he still rejected the equal brightness hypothesis by replacing it with the notion that, in all volumes of space regardless of direction, the spread of stellar luminosities is the same.

Relying on his severe mathematical training and resurrecting Gyldén's 1872 suggestion relating to this problem, Seeliger formulated an integral equation that allowed him to express the star-ratios in terms of known empirical data (principally star-counts and apparent magnitudes) for those stars brighter than magnitude 11.5 (see Eq. 2, fundamental equation of stellar statistics).

$$A_m = \int \Phi(M)dM \int D(r)dr \qquad (2)$$

This equation represents a mathematical relationship between the star-ratio (A), the luminosity law (Φ), and the density law (D). In 1912, the Swedish astronomer C. V. L. Charlier christened Seeliger's equation the "fundamental equation of stellar statistics," a name that in contemporary astronomical parlance is almost uniformly still used.[52] Assuming the general luminosity function and utilizing his three previously derived empirical laws for the star-ratio, Seeliger solved the fundamental equation for the density relationship (see Eq. 3, the density law, and Appendix II).

$$D = \lambda r^{-\gamma} \qquad (3)$$

where λ and γ are empirical constants. The result is that the density of stars (D) is entirely a function of its inverse distance from the Sun (r) proportional to the galactic latitude.

Seeliger claimed that his modified Newtonian gravitational theory implied that the mass-density of the sidereal system does not gradually thin out, but rather that it drops to zero at some *finite* distance. Although the critique of Newton's law had been theoretical, with the mathematico-statistical development of this stellar-cosmological model in 1898 Seeliger believed he had obtained an independently formulated theory whose conclusions

---

[52] C. V. L. Charlier, "Studies in Stellar Statistics – Constitution of the Milky Way," *Meddelanden Fran Lunds Astronomiska Observatorium* (Serie II), no. 8 (1912), 7 (6–63). The modern derivation of this equation can be found in most contemporary texts on galactic theory; see, for example, D. Mihalas and P. M. Routly, *Galactic Astronomy* (San Francisco: W. H. Freeman, 1968), pp. 53–5.

could be used to provide an empirical examination of his Newtonian objections. Seeliger spent the better part of the next twenty years formulating the size – and indirectly the mass-density – of his sidereal model in order to determine the total number of stars it contained. Given Seeliger's approach, the latter was possible only if one could accurately determine the star-ratio distribution for *all* stellar magnitudes. His 1898 cosmological model lacked precise data for stars fainter than magnitude 11.5. Using the fundamental equation of stellar statistics, however, Seeliger was able to express with (mathematical) confidence star-ratios in terms of empirical parameters for those stars fainter than magnitude 11.5 by assuming, as a first approximation, the universal validity of his density law and a luminosity distribution that was constant regardless of direction. From his equations he derived numerical values for both the total numbers of stars in the sidereal system and its size. In this way he hoped to verify his earlier views about the inadmissibility of the universality of Newton's gravitational law.

Although Seeliger's discourse was excessively dense and his research published almost exclusively in German journals (and therefore in the German language), his investigations slowly reached a wider audience in the non-German-speaking world. The American astronomer Simon Newcomb, in successive editions of his *Popular Astronomy,* continued to summarize for the English-speaking world – however briefly – the ideas of Seeliger, Kapteyn, and other statistical astronomers. But as Harlow Shapley revealed to Hans Kienle in 1922, "Seeliger's conclusions [on galactic theory], as given in successive editions of Newcomb-Engelmanns *Popular Astronomy* [*sic*], have not remained the same."[53] Newcomb may have been impressed with Seeliger's work and Shapley frustrated with a lack of consistency, but Seeliger was not mutually impressed with Newcomb.[54] Although Newcomb was most likely unaware of Seeliger's real feelings, various editions of Newcomb's book continued to be translated into German.[55] By

[53] H. Shapley to H. Kienle, 11 October 1922 (Shapley).
[54] H. Seeliger to M. Wolf, 17 April 1904 (Heidelberg).
[55] See H. Seeliger, "Die räumliche Verteilung der Sterne," in S. Newcomb, *Newcombe-Engelmanns populäre Astronomie* (Leipzig: Wilhelm Engelmann Verlag, 3rd ed., 1905), pp. 611–15.

1906, the astronomer and science writer Hermann Kobold published (alas in German) *Der Bau des Fixsternsystems*, easily the best and most complete nontechnical discussion of statistical cosmology to date. Slowly, Seeliger's work was being disseminated through both scientific and more popular channels.

4

KAPTEYN AND THE
DISTRIBUTION OF STARS

In their studies of the architecture of the Milky Way system, Kapteyn, Seeliger, and other statistical astronomers came to recognize the importance of three fundamental relationships: (1) the luminosity function – describing the spread of stellar brightness by magnitude class (and spectral type), (2) the density function – expressing the absolute numbers of stars per unit volume, and (3) the velocity function – a probability function yielding the actual numbers of stars as a function of their space velocities and magnitudes.

EARLY DEVELOPMENTS

Although Kapteyn's attack on the sidereal problem focused directly on all three relationships, his work should be divided roughly into two periods in which he conceptualized the problem in somewhat different terms. Up until about 1904, Kapteyn used primarily proper motion and parallax data for understanding the arrangement of the stars in space. This early work culminated in his 1902 discovery of "star-streaming," first publicly announced in 1904. The star-streaming discovery, according to which stars appear to move in two distinct and diametrically opposite directions, represented a sort of watershed in his thinking. Not only did he confirm the existence of preferential stellar motions within a dynamic Milky Way, a discovery Arthur S. Eddington later considered one of the five most important events in astronomical history during the preceding 100 years, but Kapteyn also began to realize the inadequacy of stellar motion studies alone to solve the sidereal problem. After this discovery he emphasized stellar luminosity and density studies vis-à-vis the stellar velocity relationship as a means toward unraveling the complex arrangement of the stars in space. Although the direction of his research shifted after the discovery of star-streaming, his earlier motion studies still remained an ingredient in his later luminosity and density work.

Kapteyn trained mostly in the rarified atmosphere of mathematics and physics, but upon graduating from Utrecht with a physics doctorate in 1875 he obtained a position as observer at the Leiden Observatory. Once he accepted this position, Kapteyn immediately set about learning the practical necessities of his newly adopted profession. In his new post his abilities were soon recognized, and early in 1878 Kapteyn was selected, from among four highly qualified candidates, to fill the newly instituted chair of astronomy and theoretical mechanics at the University of Groningen.[1] For his inaugural address to the university body, Kapteyn chose the topic "The parallax of the fixed stars," which suggests that he already regarded stellar distances as requisite knowledge for an understanding of the sidereal problem. Recognizing early on the importance of a broad-based set of stellar data, Kapteyn eagerly perused the extant star catalogues; but even the great *Durchmusterung* catalogues of Argelander and Schönfeld were limited to magnitudes and position, and therefore lacked the data on stellar motion that he believed so important. Kapteyn also recognized that he possessed neither the financial resources nor the proper geographical location to undertake successful observational studies. Yet if he was to explore the nature of the Milky Way, it would be necessary to acquire massive amounts of the right kind of data. As it happened, during this period David Gill (1843–1914), who was then the leader in practical astronomy and the director of the observatory at the Cape of Good Hope, was attempting to fill the hiatus created by Schönfeld's delimitation of the southern *Bonner Durchmusterung* to −22 degrees declination. Recognizing the importance of Gill's work, Kapteyn offered to help him measure the numerous photographic plates and catalogue the virtual countless number of stars. In collaboration with Gill, therefore, Kapteyn recognized a real opportunity for mutual benefit: Gill would provide the raw

[1] Among the unsuccessful candidates was Hugo von Seeliger, Kapteyn's lifelong friendly rival; see J. Schuller tot Peursum-Meijer, "De sterrenkunde voor Kapteyn (1614–1878)," in A. Blaauw et al., *Sterrenkijken Bekeken* (Groningen, 1983), 28 (7–31).

Figure 8   Jacobus Cornelius Kapteyn, ca. 1914. (Courtesy of Mount Wilson and Las Campanas Observatories, Carnegie Institution of Washington)

observational data, and Kapteyn would reduce the material for Gill and, in turn, use it in his investigation of the sidereal problem. In offering his assistance to Gill, Kapteyn wrote in 1886: "I think my enthusiasm for the matter will be equal to (say) six or seven years of such work."[2]

In the end, it took nearly thirteen years of Kapteyn's constant attention before the *Cape Photographic Durchmusterung* was completed in three volumes published between 1896 and 1900. Thus began Kapteyn's international cooperation in astronomical research. By the time the *Cape* project had gone to press, Kapteyn's laboratory had become institutionalized, and named the "Astronomical Laboratory at Groningen." Later, after Kapteyn's 1906 "Plan of Selected Areas," according to which a number of observatories would coordinate their observational work of selected stellar regions, had been adopted, Kapteyn's astronomical laboratory provided the resources for reduction and analysis of data collected worldwide. His American colleague, Frederick H. Seares, later wrote that "Kapteyn presented the unique figure of an astronomer without a telescope. More accurately, all the telescopes of the world were his."[3]

Having acquired both extensive experience and the astronomical data, during the 1890s Kapteyn produced the first of a series of papers on the nature of stellar motion, all with the purpose of eventually solving the sidereal problem. Of course, these studies were not developed in a vacuum, but relied on the earlier work of numerous nineteenth-century astronomers. As we have already seen, an understanding of the arrangement of the stars in space had been a major problem since the late eighteenth century when William Herschel enunciated his project, the "Construction of the Heavens." A central, theoretical concern was over the kind of data that could be used accurately to measure stellar distances. In the absence of other methods, Herschel had suggested that distance was proportional to stellar brightness by the principle that

[2] J. C. Kapteyn to D. Gill, 16 December 1885, reprinted in *Cape Photographic Durchmusterung* i (1896), p. xiii of the *Annals of the Cape Observatory* iii (1896) (original letter lost). Also see D. Gill to J. C. Kapteyn, 9 January 1885 and 22 January 1886 (K. A. L.) in which Gill gladly accepts Kapteyn's generous offer for assistance.

[3] F. H. Seares, "J. C. Kapteyn," *Astronomical Society of the Pacific, Proceedings*, 34 (1922), 233 (233–53).

"faintness means farness." He built his cosmology on this principle, even though he was aware, at least by the early years of the nineteenth century, of the contradiction implied by the existence of binary and multiple star systems with members of different luminosities.

As discussed earlier in Chapters 1 and 2, this anomaly motivated many nineteenth-century scientists to seek an alternative and rigorously defensible measure of stellar distances. In particular, it encouraged others to examine carefully the view that proper motions could be used as a reasonable measure for stellar distances. Drawing on these various developments, Kapteyn examined the proper motions in Auwers's catalogue and obtained a relationship correlating known (trigonometric) parallaxes with proper motions and magnitudes. Generalizing his results over large numbers of stars, he derived a statistical function that expressed "mean" distances. Published in 1901, the "mean parallax" relationship culminated not only years of careful analysis of stellar motion data, but also led directly to Kapteyn's luminosity function, which he obtained within a few months. Thus his analysis yielded a method for determining distances to stars, and it allowed him conceptually to recast his solution to the sidereal problem in terms of the luminosity function, rather than the velocity function. Although there were some minor emendations to his results, it was widely believed that his mean-parallax function was both conceptually and empirically substantially correct. In terms of the contributions of classical statistical astronomy, the mean parallax formula has been exceeded in importance only by the luminosity and density laws, and by the "fundamental equation of stellar statistics."[4]

METHODOLOGY AND THE DISCOVERY OF
STAR-STREAMING

During the 1890s, Kapteyn's investigations were based on the supposition that stellar motions were the key to understanding

---

[4] For a survey of various forms proposed for the mean parallax formula, see R. A. Robb, "The Correlation Between Absolute Magnitude, Linear Tangential Velocity, Distance, Apparent Magnitude and Proper Motion," *Royal Astronomical Society, Monthly Notices*, 97 (1936), 67–75.

the distribution of the stars. His developments were later expressed in the so-called velocity law, which attempted to relate the velocities of stars to their brightnesses. This relationship, Kapteyn argued, would provide not only an understanding of the stellar system, but would also lead to the derivation of the density and luminosity laws. The latter, it was hoped, would in turn lead to a detailed understanding of the Milky Way system. In the final analysis, however, conclusive observational evidence for his velocity relationship was entirely lacking. In the words of Eddington, of all the numbers of the famous *Groningen Publications*, a series forming "one of the most often consulted works in an astronomical library, No. 6 – the one which has never been written – [is] perhaps the most famous of them all...."[5] How could an unwritten work be so significant?

Speaking before the Amsterdam Academy of Sciences in 1895 and again in 1897, Kapteyn announced the creation of a mathematico-statistical theory expressed in the form of integral equations that related star-counts, the density function, and a Gaussian probability function of proper motions.[6] A necessary component of his theory was the standard later-nineteenth-century assumption that stellar motions are randomly distributed. With his mathematician brother, Willem Kapteyn, a professor at the University of Utrecht, as coauthor, the complete discussion of his velocity equations was published in 1900 as No. 5 in the *Groningen Publications*. In the introduction he succinctly stated their purpose:

> In what follows, an attempt will be made to deduce from the observations, what, for the sake of brevity, I will call the *law of velocities*, i.e., the law by which is defined the number of stars having a linear velocity equal to, double, triple, ..., half, a third, ... that of the solar system in space, or shorter: the law by which the frequency of a linear velocity is given as a function of its magnitude. The fundamental hypothesis on which

5   A. S. Eddington, "J. C. Kapteyn," *The Observatory*, 45 (1922), 265 (261–5).
6   J. C. Kapteyn, "Over de verdeeling der Kosmische snelheden," *Verslagen der Zittingen van de Wis- en Natuurkundige Afdeeling der Koninklijke Akademie van Weterschappen te Amsterdam*, 4 (1896), 4–18, and 6 (1898), 51–60.

this derivation rests is the following: ... The real motions of the stars are equally frequent in all directions.[7]

The observational evidence supporting the theory was earmarked for No. 6 of the *Groningen* series. Though it represented the most up-to-date views of the Kapteyns' velocity law, the theory, in Eddington's words, "turned out to be so wide of the mark that not even the beginnings of a comparison [with the observational evidence] could be made."[8] The reason for the discrepancy between theory and observation was the invalidity of the fundamental hypothesis of random motions. Although most nineteenth-century astronomers considered this hypothesis as valid a priori, in 1895 Hermann Kobold had already shown conclusively that a random distribution did not represent the observed motions of nearby stars in Auwers's catalogue.[9] Actually, the Russian astronomer Marian A. Kovalsky had already noticed in 1859 that stars in the Bradley catalogue show a preference for direction unaccountable by the parallactic effect of the solar motion.[10] Kovalsky found no logical explanation for this startling fact, but he had in fact discovered what Kapteyn was to explain forty-five years later in terms of preferential motions. Meanwhile, the anomaly between theory and evidence represented a critical problem for Kapteyn's program, because such a basic discrepancy directly affected one of his stated aims: the derivation of the density and luminosity laws from the allegedly more fundamental velocity relationship.

Despite the importance of theory in directing Kapteyn's research

---

[7] J. C. Kapteyn and W. Kapteyn, "On the distribution of cosmic velocities. Part I: Theory," *Groningen Publications*, 5 (1900), 1 (1–87).

[8] Eddington, "J. C. Kapteyn," p. 264.

[9] Although the first indication that a random distribution did not represent the observed motions of the nearby stars was noted by H. Kobold in "Ueber die Bewegung im Fixsternsystem," *Astronomische Nachrichten*, 125 (1890), 72 (65–72), his full analysis was first presented in "Untersuchungen der Eigenbewegung des Auwers-Bradley Catalogs nach der Bessel'schen Methode," *Abhandlungen der Kaiserlicher Leopoldinisch-Carolinischen Deutschen Akademic der Naturforscher*, 64 (1895), 213–365, and subsequent publications.

[10] M. A. Kovalsky, *Sur les lois de mouvement propre des etoiles du Catalogue de Bradley* (Kasan, 1859); also see O. Struve, "M. A. Kovalsky and His Work on Stellar Statistics," *Sky and Telescope*, 23 (1962), 251–2, and A. J. Szanser, "Marian Kowalski (1821–84): A Little Known Pioneer in Stellar Statistics," *Royal Astronomical Society Quarterly Journal*, 11 (1970), 343–7.

program, he claimed to be an inductivist in his scientific methodology. By this he meant primarily that one must always search
for a greater data base, and that it was from this data base, with
a properly formed conceptual framework, that hidden relationships would emerge. In reality, however, he organized the assembled data according to preconceived ideas. His velocity theory
is a good case in point. Indeed, when it became clear that the
evidence needed to support the theory was not forthcoming, he
proposed several hypotheses to explain the alleged discrepancy:
(1) The apex value was incorrect, (2) the proper motion values
were incorrect, and (3) there were preferential stellar motions.
Although he showed theoretically that the first two explanations
could account for the failure of his theory, in the end he concluded
that both the apex value and proper motions had been calculated
correctly, and he was therefore forced to conclude that stellar
motions are not randomly distributed.[11] Kapteyn's failure to harmonize observation with theory reaffirmed for him the anomalous
nature of stellar motions, and put him on the track that culminated in his discovery of the two star streams.

Kapteyn continually emphasized the need for data, data, and
more data, thus reaffirming the inductive proclivities that eventually led to his discovery. In September 1915 Kapteyn wrote
George Ellery Hale (1868–1938), director of the Mount Wilson
Observatory, "My studies have made of me more and more of
a statistician and for statistics we must have great masses of
data, of course."[12] Kapteyn's view of proper scientific methodology was to combine both deductive and inductive approaches.
Commenting on the importance of an inductive (though non-
Baconian) approach in his letter to Hale, Kapteyn illustrated his
point with his discovery of star-streaming:

> I also believe . . . that we neglect the "Art of discovery"
> too much. My impression is that we are still not suf-
> ficiently imbued with the sense of the absolute necessity
> of proceeding by induction. Deduction sets in too soon
> and too much is still expected from it. To illustrate

11 J. C. Kapteyn, "Over de verdeeling der Kosmische Snelheden," 57–8, and
    idem, "The determination of the apex of the solar motion," *Royal Academy
    of Amsterdam, Proceedings*, 2 (1900), 353–62.
12 J. C. Kapteyn to G. E. Hale, 23 September 1915 (Hale).

what I mean take the star-streams as an example.... Schönfeld was led, I think by analogy, to consider the question: May there not be a rotating motion of the Milky Way as a whole? He made the necessary computations, but found practically nothing. Other men tried a rotating motion of all the stars in orbit in the Milky Way, not necessarily all with the same period. Some, I believe, tried to adhere to a common direction of motion.... Now all this seems to me too much deductive. We began by making a wild guess, deduce its consequences and see whether it agrees with the observations. How long might we have guessed before we ... came to put the question: Are there two star streams? I blundered along for a long time in the same mistaken way, till one day I swore to go along as inductively as I could. I made drawings showing at a glance the observed data for each point of the sky. There showed very decided deviations from what was to be expected according to existing theory [i.e., random motions]. Considering these deviations as perturbations I tried to isolate these perturbations: I superimposed all the drawings belonging to Zones in which, according to existing theory, there ought to be equiformity and took averages. The result was a figure pretty well in conformity with existing theory. This drawing I then took to represent the undisturbed form and subtraction from the individual figures then gave the isolated perturbations. There showed at once a great regularity, which regularity was almost at once seen to consist in a convergence of the lines of symmetry to a single point of the sphere. From this to the recognition of two star streams.

Thus the inductive process led in a very short time to a result which others, myself included, had tried in vain to bring out in a more deductive way, for ever so long.[13]

By 1902 Kapteyn had rejected his original velocity theory and had discovered star-streaming. Finding that the stars tend to move

[13] Ibid.

in two distinct and diametrically opposite directions, Kapteyn suggested that this phenomenon resulted from two once-distinct but now intermingled populations of stars moving relative to one another.

Kapteyn first announced his new theory of stellar motions at the St. Louis World Exhibition in 1904, and again, and more important, at the 1905 meeting of the British Association for the Advancement of Science in Cape Town. At both meetings, Kapteyn emphasized that not the "slightest doubt" could be entertained concerning the reality of star-streaming. In fact, Kapteyn declared emphatically, "*all* the stars, without exception, belong to one of the two streams."[14] His overriding concern, at this point, was not to reaffirm the phenomenon, but the necessity to *confirm* the theory, namely, that there exist two independent streams of stars passing through one another in opposite directions with different mean motions relative to the Sun. In this regard, he suggested to his BAAS colleagues that radial velocity observations might prove to be the most convenient data by which to test the theory:

> I suspect that the materials for a crucial test of the whole theory by means of these radial velocities are even now on hand in the ledgers of American astronomers – alas not yet in published form. It is this fact which long restrained me from publishing anything about these systematic motions, which, in the main, have been known to me for three years [since 1902].[15]

He had in mind the Lick Observatory staff in California and particularly W. W. Campbell (1862–1938), who had been measuring radial velocities since about 1900 and possessed the data needed to confirm Kapteyn's hypothesis conclusively. Unfortunately, from Kapteyn's point of view, Campbell was less interested in verifying someone else's hypothesis (such as star-streaming) than in doing his own original work. Accordingly, during a long

---

[14] J. C. Kapteyn, "Star-streaming," *Report of the British Association for the Advancement of Science*, section A, 1905: 264 (257–65). Also see his earlier paper "Statistical Methods in Stellar Astronomy," *International Congress of Arts and Sciences, St. Louis*, 4 (1904), 412–22 (396–425).
[15] Kapteyn, "Star-streaming," p. 264.

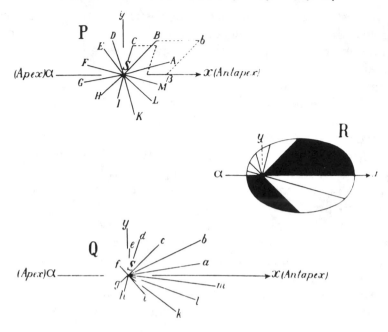

Figure 9   "Star-streaming." *P* represents the distribution of the peculiar proper motions for a group of stars located at *S*. The directions of motion are randomly distributed, whereas the length of each radial vector represents the magnitude of motion. When the observer is in motion toward the apex, the observed motions of stars as in *P* becomes *Q*. In *R* the radial vectors making angles between 0° and −60° and between 60° and 180° have been blackened. (J. C. Kapteyn, "Star-streaming," *Report of the British Association for the Advancement of Science*, section A, 1905: 257–65, p. 258)

tenure as research associate at the Mount Wilson Observatory from 1908 until his death in 1922, Kapteyn continually encouraged Hale and his associates there to further their radial velocity work with the 60-inch and later the 100-inch telescopes, in part to validate star-streaming.[16] He also strongly encouraged some of his European colleagues to continue with radial velocity investi-

---

[16]   J. C. Kapteyn to W. S. Adams, 10 December 1910, and J. C. Kapteyn to G. E. Hale, "Notes," 17 March 1918 (Hale); J. C. Kapteyn to K. Schwarzschild, 12 November 1911 (Schwarzschild).

gations. As he wrote to Karl Schwarzschild in 1912, "You cannot know how pleased I am that you are continuing with the radial velocity work."[17]

Kapteyn's discovery greatly stimulated interest in "preferential" stellar motions. He himself participated relatively little in these later developments, however, because he eventually concluded that star-streaming per se could not add directly to a detailed understanding of the architecture of the stellar system. Yet he remained keenly aware of the newer work, particularly the theoretical explanations of star-streaming offered by Schwarzschild and Eddington.[18] Occasionally, later in his career, he would even describe the sidereal problem as "the study of the arrangement of stars in space and their *systematic motion*."[19]

Although his own active research in star-streaming was diminishing, Kapteyn's investigations into the peculiar, proper, and radial motions of stars continued unabated.[20] In this regard, his research took two directions: study of the "velocity law," and of proper motions. It is true that Kapteyn continued to work on a theory of the space velocities of stars, where the frequency of velocities is a function of galactic latitude, magnitudes, spectra, and proper motions, but he would write as late as 1915 that work "on the velocity-law has been making hardly any progress."[21] His research interest in proper motions and parallaxes, however, remained intense. After all, parallaxes and proper motions were vital components of his theory of mean-parallaxes, the basis to his luminosity studies.[22]

[17] J. C. Kapteyn to K. Schwarzschild, 9 May 1912 (Schwarzschild).
[18] See Chapter 5 for a detailed discussion of these developments.
[19] J. C. Kapteyn to G. E. Hale, 17 March 1918, "Notes" (italics added); also see J. C. Kapteyn to W. S. Adams, 11 November 1912 (Hale). For the subsequent theoretical studies of star-streaming, see A. S. Eddington, "The systematic motions of the stars," *Royal Astronomical Society, Monthly Notices,* 67 (1906), 34–63; and K. Schwarzschild, "Ueber die Eigenbewegung der Fixsterne," *Nachrichten von der K. Gesellschaft der Wissenschaften zu Göttingen* (1907), 614–32. Kapteyn first became aware of Eddington's work from Gill; see D. Gill to J. C. Kapteyn, 10 November 1906 (K. A. L.).
[20] Kapteyn's interest in star-streaming always remained very strong. Indeed, in his final attempt at a model of the sidereal system, he explained star-streaming in terms of a gravitational attraction about a rotating system of stars.
[21] J. C. Kapteyn to G. E. Hale, 13 May 1915, and 6 August 1915 (Hale).
[22] Indeed, of the dozen papers Kapteyn published after 1915, six deal with proper motions and parallaxes. For a short time, there was some question

During the war years, Kapteyn and Hale exchanged many letters discussing scientific method and proper approaches to scientific work. Although Kapteyn's research was empirically grounded, his approach often possessed a noticeable mathematical flavor. Certainly he recognized that for many types of investigations his inductive "method" was only partially useful. In his September 1915 letter to Hale, in which he explained how he had discovered star-streaming, Kapteyn noted:

> From the standpoint of a finished theory [of star-streaming], the [inductive] method must be conceded to be very little rigorous. I have a notion that it will be so in most cases. Therefore it has to be supplemented by a deductive treatment of the problem: given that there are two streams, show rigorously that the observations are well represented. I think that it is on account of the mathematical rigor that the latter problem (the deductive one) appeals to so many even of the very best men. . . .[23]

Among the "very best men," Kapteyn included Schwarzschild and Eddington, both of whom were mathematically rigorous in their work on star-streaming.

Kapteyn concluded this letter to Hale by noting that "the inductive part of the investigation, with all its defects, is incomparably the more important part of research. The deductive problem can be solved by any well skilled mathematician." Here Kapteyn was confusing mathematical deduction with hypothesis formation and the deduction of tentative claims based on the assumptions and principles of the relevant body of knowledge. On precisely this point, Hale responded to Kapteyn's "most interesting letter of September 23," and gently corrected Kapteyn

---

about the validity of Kapteyn's parallaxes. Around 1917 W. S. Adams and G. B. Strömberg, at Mount Wilson, claimed that Kapteyn's proper motions were erroneous. They based their work on the incorrect assumption that there were no giant and dwarf stars, thus suggesting that the distances (mean-parallaxes) were much closer. On this episode, see J. C. Kapteyn to G. E. Hale, 23 May 1918, 6 October 1918, 7 October 1918, and 1 December 1918, and G. E. Hale to J. C. Kapteyn, 17 August 1918; and H. Shapley to G. E. Hale, 12 September 1920 (Hale).

23 J. C. Kapteyn to G. E. Hale, 23 September 1915 (Hale).

while attempting to convince him of the fundamental significance of "framing hypotheses":

> The account of your discovery of the two star streams is a splendid illustration of the value of the inductive method, which doubtless serves best in a large class of investigations. And yet I cannot help feeling that in many other cases a combination of deduction and induction is more likely to be successful. In fact, I constantly find myself instinctively framing hypotheses as guides to research, ... and therefore endeavoring to construct multiple hypotheses to account for obscure phenomena. Each hypothesis suggests the application of a series of criteria, and it usually becomes possible to eliminate some of them very soon.... In my experience, therefore, deductive methods are almost invariably applied. Frequently they are the merest guesses, without a substantial theoretical basis. But each suggests experiments or tests which would hardly occur to me otherwise, and thus almost any hypothesis may prove useful.[24]

Continuing, Hale showed that Kapteyn actually framed hypotheses, and in fact had done so in both his star-streaming discovery and his research on stellar velocities and movements:

> Is it not true that you also employ deductive methods in many instances? Sometimes they may enter only tacitly, but nevertheless play an important part in your investigations. In fact, on a basis of pure induction your imagination would have little opportunity to serve as a guide.... Your new result regarding the star streams is of the greatest interest and importance, especially as it offers a much more satisfactory picture of the star system than the separate streams afforded.... Is it even possible that you would not have undertaken the

---

[24] G. E. Hale to J. C. Kapteyn, 4 November 1915 (Hale). In their ensuing correspondence on scientific method, Hale gently recommended that Kapteyn read a number of philosophical works on the subject; G. E. Hale to J. C. Kapteyn, 22 September 1915, 18 October 1915, and 9 March 1916 (Hale).

investigation which led to the discovery of the star streams if you had not been started in this direction by an (incorrect) hypothesis?[25]

For our purposes, it is perhaps most important to note that Kapteyn eventually agreed with Hale that hypotheses are the essential guiding heuristics needed to make sense of disparate phenomena. In response to Hale's reply, Kapteyn wrote: "After all I think that our difference is more a difference of degree. One cannot pass from the observations to the laws underlying them, than by making certain jumps, certain hypotheses and applying these to the observations in hand, and modifying them till they fit."[26]

DISTRIBUTION OF STARS

Frustrated in his hope of deriving the velocity law because of the phenomenon of star-streaming, Kapteyn, although using his original data and his earlier work with the mean-parallax relationship, increasingly turned his attention to a more direct approach to understanding the structure of the sidereal universe. Using the basic data he had worked with for so many years, Kapteyn eventually obtained solutions for the luminosity and density functions, which describe, respectively, the spread of stellar brightnesses from the faintest stars (in absolute magnitude) to the brightest, and the density distribution of stars in terms of distances from the solar region. Thus, even though his efforts at detailing the Milky Way were temporarily thwarted by the failure of his velocity theory, he continued to use both the observational data and the results of his earlier work in order to understand the larger sidereal system. Indeed, at the very end of his career, Kapteyn returned to these early efforts and developed a dynamic model of the stellar system, in which he combined both his stellar motion studies and the more fruitful results from his luminosity and density research.

Although Seeliger was the first theoretical astronomer to provide

[25] G. E. Hale to J. C. Kapteyn, 4 November 1915 (Hale).
[26] J. C. Kapteyn to G. E. Hale, 26 March 1916 (Hale). In his letter, Kapteyn noted that he had spoken extensively on the matter of method with "our philosopher [G.] Heymans and with [Paul] Ehrenfest, [H. A.] Lorentz's successor."

a solution (in 1898) to the sidereal problem, his analysis was mathematically very abstract and it was not based rigorously on all the available observational data. Kapteyn had also been working on the sidereal problem, but he was in greater command of all the empirical data, particularly proper motions and parallaxes. Both Seeliger and Kapteyn recognized the essential importance of relating the star-counts, the density distribution, and the luminosity functions; they further realized that an accurate representation of the arrangement of the stars in space ($D(r)$ in Seeliger's terms) would require an exact understanding of the luminosity relationship. Kapteyn had already concluded during the 1890s that a precise description of stellar distances could be determined from data obtained from stellar motions and stellar brightnesses. By 1901 he had developed a *numerical* technique for obtaining the luminosity function that related the magnitudes of stars to their motions, and hence their mean distances. Kapteyn's solution, however, was more clearly comprehensible and far less abstract than Seeliger's. Partly for this reason, Kapteyn's work became more widely known than Seeliger's highly original investigations.[27]

Although Kapteyn's procedure was first published in 1901 in his classic paper "On the luminosity of the fixed stars,"[28] all the subsequent analyses of the sidereal problem that he published during the remaining years of his life followed the method outlined in this earlier work. Briefly, his method entailed placing the catalogued stars in cells corresponding to their apparent magnitude and probable proper motion (see Figure 10).[29] Utilizing his

[27] See H. Shapley to H. Kienle, 11 October 1922 (Shapley). Shapley noted that the mathematical abstractness of Seeliger's work hindered many astronomers from better understanding Seeliger's contributions. Also see H. Kienle to H. Shapley, 21 September 1922 (Shapley).

[28] J. C. Kapteyn, "On the luminosity of the fixed stars," *Koninklijke Akademie van Wetenschappen te Amsterdam. Proceedings of the Section of Sciences*, 3 (1901), 658–89; reprinted with slight corrections in *Groningen Publications*, no. 11 (1902), 3–32.

[29] During 1863–8 the Italian astronomer Angelo Secchi had examined 4,000 stars and classified them into four types (I, II, IIIa, and IIIb, according to H. C. Vogel) corresponding to their apparent color (temperature). Kapteyn used only stars with Secchi spectral types I and II, because these could more easily be derived from available star-counts. The present-day scheme of classification – O, B, A, F, G, K, M, N – corresponds to Secchi's scheme: I – B, A; II – F, G, K; IIIa – M; IIIb – N.

μ	Mean μ	1.5—2.5	2.5—3.5	3.5—4.5	4.5—5.5	5.5—6.5	6.5—7.5	7.5—8.5	8.5—9.5
		2.1	3.1	4.1	5.1	6.1	7.1	8.1	9.1
0'.000—0".009	0".005	6	5	22	90	343	1504	9010	38257
.010— .019	.015	4	15	52	194	638	1896	7313	28184
.020 .029	.025	1	10	41	177	595	1910	5882	21225
.030— .039	.035	3	16	27	188	542	1910	4768	15809
.040— .049	.045	3	12	27	93	461	1490	3877	12045
.050— .059	.055	5	13	22	86	357	1249	3342	9184
.060— .069	.065	1	4	25	77	252	963	2540	6775
.070— .079	.075	2	2	18	71	247	752	1827	4968
.080— .089	.085		5	25	45	209	692	1337	3614
.090— .099	.095	1	9	17	54	200	646	802	2409
.100— .149	.125	5	10	57	152	424	963	1649	4373
.150— .199	.175	5	9	34	79	181	420	1070	2033
.200— .299	.25	5	11	32	73	200	315	670	919
.300— .399	.35	1	3	17	43	105	75	178	422
.400— .499	.45		5	12	18	34	121	133	196
.500— .599	.55	1	3	8	7	24	38	53	60
.600— .699	.65	2		2	7	5	24	32	36
.700— .799	.75			9	5	10	17	27	26
.800— .899	.85	1		3	2		12	13	16
0.900— 0.999	0.95				2		6	9	10
1.000— 1.199	1.1		1	3	5		7.5	9	10
1.200— 1.399	1.3			3	2	5	12.0	9	7.6
1.400— 1.599	1.5				2		4 5	4.4	6.0
1.600— 1.799	1.7						1.5	4 4	1.5
1.800— 1.999	1.9			2	2		1 5	0.0	1.5
2.000— 2.999	2.5		1				6.0	8.9	6.0
3.000— 3.999	3.5					5	3 0	4.5	6.0
4.000— 4.999	4.5				2		1.5		3.0
5.000— 5.999	5.5					5		4.5	
6.000— 6.999	6.5						1.5		4 5
7.000— 7.999	7.5								
Total		46	134	458	1476	4842	15 042	44 576	150 607

Figure 10   Apparent magnitude and proper motions table. (This and the following tables are from J. C. Kapteyn, "On the luminosity of the fixed stars," *Groningen Publications*, no. 11 (1902), 8, 10, 11 [3–32])

Limits of π		fraction of the whole	Number.
0″00000 and 0″00100		0.001	0
00100 ,, 00158		.004	2
00158 ,, 00251		.028	13
00251 ,, 00398		.097	45
00398 ,, 00631		.209	96
00631 ,, 0100		.275	127
0100 ,, 0158		.226	104
0158 ,, 0251		.116	54
0251 ,, 0398		.0358	16.5
0398 ,, 0631		.0068	3.1
> 0″0631		.0009	0.4
		1.000	461

Figure 11 Mean-parallax distribution.

"mean parallax" formula, which expressed a dependency between calculated parallaxes on the one hand, and apparent magnitudes and proper motions on the other, Kapteyn next calculated the mean-parallaxes corresponding to each cell. The stars in each cell are not, however, all found at the same parallax distance, but rather they are distributed about some mean as expressed by the mean parallax formula. Thus, for example, for the 461 stars with p.m. = 0.045 and app. mag. = 6.1, the mean-parallax is 0.0102. These results represent only *mean-parallaxes*. In actuality the stars would be distributed according to the laws of probability defined for purposes of analysis by some Gaussian function. The exact shape of the Gaussian curve was derived from a determination of the spread of fifty-eight particular stars, with precisely known

π	1/π	−0.9	0.1	1.1	2.1	3.1	4.1	5.1	6.1	7.1	8.1	9.1	Vol.
0".00000—0".00100	0".00118				0.4	0.5	3.0	18.0	84.0	440.0	3016.0	15363.8	
.00100—	.00187			1	1.0	2.0	8.0	34.0	152.0	667.0	3658.0	16197.0	3 140 000
.00158—	.00296				1.9	2.7	15.2	76.0	306.0	1216.0	5419.0	21953.0	788 000
.00251—	.00469				2.9	7.9	29.1	141.0	531.0	1942.0	6912.0	25578.0	198 000
.00398—	.00743				3.8	14.9	48.7	211.0	770.8	2642.0	7880.0	26044.0	49 700
.00631—	.0118	1	1	1	5.1	20.2	64.7	255.5	901.0	2899.7	7352.0	20977.0	12 500
.0100—	.0187				6.1	23.7	74.6	248.8	833.5	2403.2	5278.5	14067.5	3 140
.0158—	.0296			2	7.3	21.8	74.5	209.4	614.3	1556.9	3003.9	6789.5	788
.0251—	.0469			2	6.5	17.6	59.2	142.4	367.5	780.5	1338.0	2568.4	198
.0398—	.0743		1	2	5.2	11.5	40.1	78.8	177.2	322.5	496.4	792.1	49.7
.0631—	.118		1	1	3.3	6.4	23.3	37.7	69.1	114.6	157.3	207.9	12.5
.0100— .158	.204		1	1	1.6	3.0	11.4	15.3	22.9	39.2	45.8	50.0	3.14
>.158		1	1	2	0.9	1.8	6.2	8.1	12.7	18.4	19.6	18.8	1.05
		2	5	10	46.0	134.0	458.0	1476.0	4842.0	15042.0	44576.5	150607.0	

Figure 12 Luminosity table.

proper motions, apparent magnitudes, and measured parallaxes. Using this computed probability distribution, Kapteyn calculated the spread of parallaxes for the stars within each cell. Thus, for example, the 461 stars are distributed about their mean as shown in Figure 11. The limiting characteristics of the distributed stars correspond to spherical concentric shells about the Sun. The first shell represents a distance of ten parsecs (Kapteyn's unit distance); all subsequent shells represent distances corresponding to an increase of one apparent magnitude. Because the stars in each cell are distributed across all shells, Kapteyn, after deriving similar results for each cell of Figure 10, produced a two-dimensional table (see Figure 12), in which the catalogued stars were distributed by magnitude class and mean distances. The magnitudes were normalized by conversion to their absolute magnitude by means of the magnitude–distance relationship. This directly yielded the spread of magnitudes or luminosities (as total illuminating power) of all stars. Finally, Kapteyn reduced his tabular representation to an analytic form.

Others, such as Hugo Gyldén, G. V. Schiaparelli, and particularly Seeliger, had noted the importance of using a Gaussian function to represent the luminosities, but Kapteyn alone succeeded in actually deriving a workable, and defensible, function. Kapteyn thus introduced the term "luminosity-curve" into astronomical parlance as "the curve which for every absolute magnitude gives the number of stars per unit of volume."[30] Because his luminosity table expressed distances from the Sun to the stars of various magnitudes, it was a simple matter of dividing the numbers of stars within each shell by the shell's volume to calculate the relative density. Thus as a by-product, Kapteyn's procedure also yielded the density distribution of stars in the local solar neighborhood. Because only stars brighter than magnitude 9.5 were used, the density relationship was tentative at best. Nevertheless, a first, provisional solution was now available.

[30] Kapteyn, "On the luminosity of the fixed stars," p. 670 (reprint p. 14).

5
STATISTICAL ASTRONOMY AS
A RESEARCH PROGRAM,
1900–1915

In this period, Kapteyn and Seeliger maintained their roles as leaders in what promised to become an extremely fruitful research field. Others, particularly van Rhijn, Schwarzschild, Eddington, and Charlier, began to make significant contributions after about 1910, but Kapteyn and Seeliger continued to define and clarify many of the major research problems dominating statistical cosmology from the beginning of their statistical studies in the 1890s until the early 1920s.

Not only were problems of substance explored, but new methods were also developed, all of which provided grist for the mills of numerous statistical astronomers. Accordingly, after about 1900 research in statistical astronomy was further characterized by investigations of the correct analytical form of the various relationships, particularly the laws of density and luminosity, in addition to that of velocity, but also the mean-parallax and star-count functions. When Seeliger's paper on the density function appeared in 1898 and Kapteyn's luminosity paper appeared in 1901, the possibilities inherent in a rigorous approach to statistical investigations were immediately improved. This was the beginning, however, more than the end. Kapteyn himself considered his 1901 paper as providing only a first approximation to the sidereal question, a problem that, in his opinion, "must be solved by successive approximations." This view was shared independently by Seeliger, who attacked the problem at least four more times by 1920.

In this work, they were utterly convinced that an understanding of these relationships would yield universal laws – not just statistical relationships, but laws of nature. After the derivation of the luminosity function in 1920, Kapteyn expressed it this way: "It is difficult to avoid the conclusion that we have here to do with a law of nature, a law which plays a dominant part in the most diverse natural phenomena."[1]

ASSUMPTIONS AND RESEARCH PROBLEMS, CIRCA 1900

It has frequently been observed that the significance of a research program lies not so much in the number of questions that are answered, but in the number of areas for additional research that open up as a result of the work. As a "first attempt" at the solution to the sidereal problem, Kapteyn's and Seeliger's studies made several critical assumptions that in the following years created extremely fruitful research areas for statistical astronomers by defining key problems. To varying degrees, their initial research had assumed: (1) negligible light absorption; (2) a Sun-centered stellar system; (3) a luminosity curve uniform throughout the entire stellar system (i.e., independent of distance and direction from the Sun); (4) a luminosity curve distributed according to Secchi's type I and II stellar spectra; (5) a density relationship independent of galactic longitude and latitude; and (6) true parallaxes of stars distributed about their mean in a Gaussian symmetric form.

Interstellar absorption

The question of the transparency of space had been discussed by many nineteenth-century astronomers, including William Herschel, W. Olbers, F. G. W. Struve, Kapteyn, and Seeliger. Kapteyn recognized that the existence of an interstellar absorbing medium could seriously alter the form of the luminosity curve and thus fundamentally change the parameters describing his stellar system. During his vacation months from 1908 until the outbreak of the First World War, while Kapteyn held a research associate position at Mount Wilson, he remained only peripherally involved in this research, but he encouraged others on the Mount Wilson

[1] J. C. Kapteyn and P. van Rhijn, "On the Distribution of the Stars in Space Especially in the High Galactic Latitudes," *Mount Wilson Observatory, Contributions*, no. 188 (1920), reprinted in *Astrophysical Journal*, lii (1920), 33 (23–38).

staff to study the problem. By 1915, Kapteyn himself had submitted three papers dealing with it.

Between the time of his earlier star-streaming work and 1915, Kapteyn's view on the question of absorption changed enormously. He – and others working on the problem – began to realize that it involved not only general spatial absorption, but also selective absorption (dependency on wavelength), "scattering in space" of stellar light, and photographic film absorption. During most of his early work on the sidereal problem, Kapteyn had made the (reasonable) assumption that stellar absorption was negligible, for without additional empirical justification any other assumption would have been entirely ad hoc. Under this assumption, his derived density distribution function gave him a maximum value in the solar region, justifying a Sun-centered model. The question, of course, was whether in fact the density distribution was real or merely apparent. And the answer to this question depended in large measure on the existence of interstellar absorption.[2]

Kapteyn's interest in the absorption question was heightened when around 1904 the American astronomer George C. Comstock examined the problem in a somewhat superficial analysis using techniques and data derived from Kapteyn's 1901 work. Comstock used the outmoded assumption of equal intrinsic stellar brightness, which, when coupled with Kapteyn's results for stellar distances, led him to assert an absorbing coefficient of 17 magnitudes per kiloparsec (mag/kpc). Kapteyn was quick to point out that Comstock's results led to entirely erroneous density distributions within as little as one kiloparsec of the Sun. But, spurred on by Comstock's work, Kapteyn analyzed the general absorption question, and found that for various absorption coefficient values, a nearly constant stellar density distribution could be obtained with an absorption of only 1.6 mag/kpc. This result had the advantage that it rationalized away a Sun-centered cosmology, implying a "uniform" universe without a recognizable center.[3]

[2]   For a useful analysis of the absorption question, see D. Seeley's dissertation; also see D. Seeley and R. Berendzen, "The Development of Research in Interstellar Absorption, c. 1900–1930," *Journal for the History of Astronomy*, 3 (1972), 52–64 and 75–86.

[3]   G. C. Comstock, "Provisional Results of an Examination of the Proper Motions of Certain Faint Stars," *Astronomical Journal*, 24 (1904), 43–9; and J. C. Kapteyn, "Remarks on the Determination of the Number and Mean

His 1904 results were tentative at best, and Kapteyn admitted as much. Other methods, therefore, would have to be utilized. Because reliable, empirical results remained scanty, within the next few years a critical assessment of instruments and techniques led Kapteyn and others again to question their absorption co-efficient results. Writing to the Mount Wilson astronomer Harold D. Babcock early in 1909, Kapteyn suggested:

> As the silver grain of the film must cause at least some scattering of light, it seems almost necessary, that the transmitted light must be somewhat reddish – on the same ground on which the scattering in space must give a reddish colour to the stars. If so, then the effect must be much diminished and may even be reversed. . . . If you had hundreds of bright and faint stars, you could even probably determine from your plates, both the amount of space – and of film – absorption.[4]

He therefore proposed to search for reddening of light owing to a scattering dependent on wavelength. Because the eye is less sensitive to blue light than it is a photographic plate, Kapteyn suggested comparing the magnitudes of stars obtained visually with those obtained photographically. The result revealed an absorption of 0.3 mag/kpc.[5]

Kapteyn did not return to serious research on absorption until his summer tenure at Mount Wilson in 1913. Though his results during these few summer months were preliminary, Kapteyn predicted that researchers would find significant general absorption, and not primarily selective absorption, owing to interstellar material. Walter S. Adams (1876–1956), a leading spectroscopist and second-in-command at Mount Wilson, and Frederick H. Seares (1873–1964) and Arnold Kohlschütter (1870–1942), permanent staff observers at Mount Wilson, continued to study the general spatial absorption predicted by Kapteyn, and soon obtained encouraging results. In order to minimize photographic errors in

Parallax of Stars of Different Magnitude and the Absorption of Light in Space," *Astronomical Journal*, 24 (1904), 115–23.
[4] J. C. Kapteyn to Harold D. Babcock, 8 November 1909 (K. A. L.).
[5] J. C. Kapteyn, "On the Absorption of Light in Space," *Astrophysical Journal*, 30 (1909), 284–317.

their search for the effects of interstellar absorption, Kapteyn had unwittingly suggested to them that they place the spectra of both high and low proper motion stars of the same spectral class on the same photographic plate.[6] As Hale reported to Kapteyn in January 1914: "Adams' pairs of spectra . . . are the most striking ocular evidence of general absorption yet available. . . . With the new material now available I hope you will be in a position to complete your paper on the general question of space absorption, as the 'psychological moment' for its publication seems to have arrived."[7] By the "psychological moment," Hale had in mind this work of Adams, Seares, and Kohlschütter that suggested some evidence for general absorption. Replying to Hale's letter, Kapteyn also emphasized the importance of wavelength-dependent absorption vis-à-vis general absorption: "I feel quite elated about the turn that the absorption of space work is taking, in particular if it really turns out that there is an influence on the hydrogen [spectral] lines. It is almost too beautiful to be true."[8] The most spectacular result to emerge from the work of the Mount Wilson astronomers indicated that, regardless of whether there exists interstellar absorption or not, there is an empirically verified absorption of star-light passing through interstellar material. In writing to Hale in March of 1914, Kapteyn remarked: "There is nothing in which I am so interested as in what you and Adams write about the influence of distance on the hydrogen lines. If we could but find some lines altogether due to gaseous matter in space! It should mean I think almost a revelation in astronomy."[9] Although Adams initially agreed with this interpretation, he was

[6] For a discussion of this episode in the broader context of stellar evolution, see David DeVorkin, "Stellar Evolution and the Origin of the Hertzsprung–Russell Diagram," in Owen Gingerich, ed., *Astrophysics and Twentieth-Century Astronomy to 1950: Part A* (Cambridge: Cambridge University Press, 1984), pp. 103–4 (90–108).

[7] G. E. Hale to J. C. Kapteyn, 6 January 1914 (Hale). At first the tentative results of Adams and Seares were not published, but communicated by Hale to Kapteyn (who was living in Holland) so that the latter could examine the material and offer his opinion of the results; J. C. Kapteyn to G. E. Hale, 11 September 1913, and G. E. Hale to J. C. Kapteyn, 6 January 1914 (Hale).

[8] J. C. Kapteyn to G. E. Hale, 12 February 1914 (Hale).

[9] G. E. Hale to J. C. Kapteyn, 29 May 1914 (Hale). On the recognition of selective absorption as significant, see also J. C. Kapteyn to G. E. Hale, 6 April 1914, and G. E. Hale to J. C. Kapteyn, 11 May 1914 (Hale).

aware that the spectral differences could also be due to luminosity differences. After the Danish astronomer Ejnar Hertzsprung (1873–1967) wrote Adams reminding him of this possibility, Adams concluded that luminosity and not absorption was the explanation of their earlier discovery. Consequently, by May of 1914 Hale had reversed his earlier enthusiasm for the current work of the Mount Wilson researchers. "It does not now seem probable," he wrote to Kapteyn, "that the hydrogen absorption occurs in space, as the lines appear to be displaced equally with other stellar lines. This . . . indicates that the star itself is responsible for the changes of relative intensity."[10]

Although Kapteyn continued slowly to relinquish his earlier view of substantial spatial absorption, he still hoped conclusive evidence would eventually appear. Unfortunately, this was to remain an illusive quest for another decade, when empirical evidence of line absorption was at last discovered. The definitive work on general absorption was not done until 1929–30 by Robert Trumpler. Until then, other approaches to the various absorption questions were tried.

Among various researchers investigating interstellar absorption, the astronomers at Mount Wilson remained intensely interested in the question. Just prior to receiving his Ph.D. from Princeton in 1913, the young American astronomer Harlow Shapley (1885–1972) obtained an appointment from Hale as an assistant at the Mount Wilson Solar Observatory.[11] Soon after he arrived at the Pasadena facilities in January 1914, Hale assigned Shapley the task of investigating globular clusters. After extensive studies of the stars in many of the clusters, Shapley concluded that the clusters must be between 10 and 100 kiloparsecs from the solar region. At these distances, if interstellar absorption existed, even in the small amounts Kapteyn proposed, then all the cluster stars should be reddened. In fact, the colors of stars ranged from red to blue, hardly an encouraging result for Kapteyn's program. Interstellar space, concluded Shapley by late 1915, must be transparent: Kapteyn's earlier value "must be from ten to a hundred times too large . . . and the absorption in our immediate region

10 J. C. Kapteyn to G. E. Hale, 29 March 1914 (Hale).
11 H. Shapley to G. E. Hale, 14 November 1912; G. E. Hale to H. Shapley, 26 December 1912 (Hale).

of the stellar system must be entirely negligible."[12] Even though Kapteyn personally wrote Shapley acknowledging the latter's cluster results as both "fine and surprising," he also expressed his disappointment that an empirically verifiable interstellar absorption had not yet been obtained.[13]

Sun-centered stellar system

Lacking a plausible alternative, statistical astronomers had generally assumed that the Solar System was centrally located in the Universe. To be sure, the nature of this assertion made many feel increasingly uneasy; yet, as a working hypothesis, it was simple and reasonably defensible. As we have indicated, by 1915 studies had suggested a lack of an absorbing medium, and the results supported a maximum density distribution within the immediate solar neighborhood. In the September 1915 letter to Hale quoted earlier, in which Kapteyn discoursed on method and on star-streaming, he had concluded that the direction and velocity of the star-streams also gave credence to the privileged solar position:

> The stream velocity increases with decreasing distance from the sun. The result seems to me to be well established. One of the somewhat startling consequences is, that we have to admit that our solar system must be in or near the center of the universe, or at least to some local center. Twenty years ago this would have made me very sceptical ... Now it is not so. Seeliger, Schwarzschild, Eddington and myself have found that the number of stars is greater near the sun. I have sometimes felt uneasy in my mind about this result, because in its derivation the consideration of the scattering of light in space has been neglected. Still it appears more

---

[12] H. Shapley to F. Moulton, 7 January 1916 (Shapley). For a review of Shapley's cluster color studies work, see his "Studies Based on the Colors and Magnitudes in Stellar Clusters. First Part: The General Problem of Clusters; Second Part: Thirteen Hundred Stars in the Hercules Cluster (Messier 13)," *Astrophysical Journal*, 40 (1917), 118–40.

[13] J. C. Kapteyn to H. Shapley, 17 August 1915 (Shapley). For similar remarks about Shapley's innovative work, also see J. C. Kapteyn to G. E. Hale, 7 March 1915 (Hale).

and more that the scattering must be too small, and also somewhat different in character from what would explain the change in apparent density. The change is therefore pretty surely real. Here then we have another, really very strong indication that the position of our solar system is nearly central.[14]

His letter to Hale confirmed privately only what he had, for all intents and purposes, already committed himself to publicly as early as 1901: that the local solar neighborhood seemed centrally located within the larger sidereal system. Although he included the usual qualification that the density distribution remained provisional, even his (1908) solution of the sidereal problem confirmed the earlier general solution (1901) of a Sun-centered stellar system.

Uniform luminosity function

Kapteyn, even more so than Seeliger and certainly more than most others, had control of the massive amounts of data needed to understand the luminosity frequency relationship. His numerical scheme developed first in 1901 was particularly well suited to continued exploration of this problem. As Kapteyn expressed it to Hale in 1908:

> I am just now heads over ears working out the problem of the thinning out of the stars as we recede from the solar system. In [Groningen] Publ. 11 I think I have arrived at what I consider a fairly trustworthy determination of the luminosity "curve," but, as mentioned there, the law of the densities here derived could only be considered as a very provisional and rough one.[15]

Following the results of 1901, Kapteyn's 1908 work had also assumed negligible absorption.

[14] J. C. Kapteyn to G. E. Hale, 23 September 1915 (Hale).
[15] J. C. Kapteyn to G. E. Hale, 18 February 1908 (Hale). J. C. Kapteyn, "On the Number of Stars of Determined Magnitude and Determined Galactic Latitudes," *Groningen Publications*, 18 (1908), 31 pp.; and idem, "On the Mean Star-Density at Different Distances from the Solar System," *Koninklijke Akademie van wetenschappen te Amsterdam. Proceedings of the Section of Sciences*, 10 (1908), 626–35.

It could be assumed that all regions of space, regardless of galactic position or distance from the Sun, exhibit precisely the same *distribution* of luminosities within the same unit volume. The assumption that "the luminosity-curve is the same for different distances from the sun" was therefore theoretically independent of the question of interstellar absorption.[16] However, because the derivation of Kapteyn's luminosity function required empirical knowledge of parallaxes and proper motion, as well as of stellar luminosities, the function could actually be determined only for the local solar neighborhood. Although he devoted a considerable amount of time to analyzing the luminosity function, the precise analytic shape of the luminosity curve continued to remain a difficult problem. Nevertheless, Kapteyn hoped additional information could be obtained that would extend data particularly for stars of increasingly fainter magnitudes and of all spectral types. Already as early as 1909, for example, he argued vehemently on theoretical and practical grounds for the construction of the 100-inch telescope needed for solution of the sidereal problem: "[E]very inch of aperture gained promises a great advance in astronomical possibilities.... For if we can determine both the frequency law [luminosity law] and the absorption of space, the determination of the real arrangement of stars in space very probably will not present any very serious obstacles."[17]

Luminosity spectral relationship

Not only were stellar magnitudes directly involved, but the spectral nature of stars also complicated matters considerably. Kapteyn had already noted a correlation between spectral type and proper motions in a 1892 study,[18] and in his 1901 studies of the stellar system Kapteyn noted a relationship between spectra and luminosities. But the revolutionizing developments in spectral

[16] See J. C. Kapteyn, "On the Luminosity of the Fixed Stars," *Koninklijke Akademie van Wetenschappen te Amsterdam. Proceedings of the Section of Sciences*, 3 (1901), 670–4 (658–89); reprinted with slight corrections in *Groningen Publications*, no. 11 (1902), 3–32.

[17] J. C. Kapteyn to G. E. Hale, 21 September 1909 (Hale).

[18] See the discussion of this issue in Chapter 2.

classification by Annie Jump Cannon, Henry Norris Russell, and Ejnar Hertzsprung early in the century urged Kapteyn and Seeliger to emphasize the importance of detailed spectral studies. This presented a real problem – and "a new and most powerful aid" – to astronomical investigations.[19] Because the spectral scheme has a real physical meaning, it is not simply an ad hoc typology. As Eddington put it:

> This spectral classification may be regarded as primarily an attempt to arrange the stars according to the stage of evolution that has been reached; and it is believed that the stages are passed through in the order of the letters [B A F G K M]. . . . This series may also . . . be regarded as the order of stellar temperatures, B being the hottest and M the coolest stars.[20]

Writing to J. A. Parkhurst at the Yerkes Observatory as early as 1909, Kapteyn argued that "for questions of distribution of stars in space it becomes more and more evident how desirable it will be to carry the whole thing through for each spectral class *separately*."[21] In reply, Parkhurst concurred: "I fully agree with you as in the importance of using the spectral classification in investigating stellar distribution."[22]

Kapteyn remained adamantly committed to this project for many years. In personal correspondence to Schwarzschild in 1910, Kapteyn noted perhaps somewhat overly optimistically: "I have undertaken another attack on the sidereal system; there is now over the entire sky enough data, and also we would like to make a study of the later spectral types (A, A1, . . .) and how they are distributed."[23] Writing to Adams in 1912 concerning the crucial importance of the 100-inch telescope for spectral work, Kapteyn continued:

---

[19] A. S. Eddington, "Stellar Distributions and Movements," *Observatory*, 34 (1911), 358 (355–9).

[20] A. S. Eddington, "The Distribution of the Spectral Classes of Stars," *Observatory*, 36 (1913), 467 (467–71).

[21] J. C. Kapteyn to J. A. Parkhurst, 5 April 1909 (Yerkes).

[22] J. A. Parkhurst to J. C. Kapteyn, 20 April 1909 (Yerkes).

[23] J. C. Kapteyn to K. Schwarzschild, 22 October 1910 (Schwarzschild). Kapteyn's 1910 attack on the sidereal system was published as "The Luminosity Curve," *Astronomische Nachrichten*, 183 (1910), 313–32.

Now that we begin to know something of the luminosity curve of stars of one particular class of spectrum –
I have already derived such a curve for the Helium [or
B] stars and will shortly try to do so for the A stars. . . .
Before we can really . . . attack the question of the
arrangement of stars separately for the different spectra, such a knowledge [of the colors] seems absolutely
necessary. In my mind, the most important problem in
sidereal astronomy would be: the study of the arrangement of stars in space (including star streams) *separately*
for stars of different spectral type. . . .[24]

A few years later, Kapteyn outlined for Hale the progress he and
others were making concerning this critical question.

[W]e can find the distributions in space of nearly all
the Helium stars [and] . . . there is a gradual transition
in every direction from the Helium stars to the other
types. . . . All this finished I will have to come to the A
stars, which in the main I find to behave like the Helium stars. If I finish them too I think I may hope to
solve the many riddles that remain for the rest."[25]

Density galactic relationship

Stars are not only distributed by spectral type, but they are also
distributed by galactic position. Though the Herschels, Struve,
and many others had noted the dependency of the number of

[24] J. C. Kapteyn to W. S. Adams, 11 November 1912 (Hale).
[25] J. C. Kapteyn to G. E. Hale, 26 March 1916 (Hale). The Swedish astronomer C. V. L. Charlier reformulated the statistics of stellar astronomy by
considering the mathematics of statistical errors. Recognizing inherent problems in any discussion of the *statistics* of a wide range of data (stars, in this
case), Charlier showed that the situation could be improved greatly if star-
counts were organized not only according to apparent magnitudes, but also
according to spectral types. Doing so led to his conclusion that the B-type
stars are distributed eccentrically mostly in the plane of the Milky Way. It
was this result that later convinced Charlier to accept Shapley's argument
that the Solar System is not centered in our galactic system. See the discussion in Chapter 8.

stars (stellar density) on galactic latitude, it was Seeliger who, in the 1880s, first rigorously demonstrated this fact.[26] In later studies Kapteyn recognized Seeliger's work on this point, and noted that the luminosity curve, being a distribution function and not an absolute measure, is independent of both galactic longitude and latitude. On the other hand, the density function is, in the final analysis, an absolute measure of the numbers of stars per unit volume of space. Therefore, it would depend not only on the distance from the solar region (assuming the stellar system is Sun-centered), but also on galactic longitude and latitude.[27]

Gaussian stellar parallaxes

With Adams's earlier work (1914) on space absorption using the spectra of low and high proper motion stars, he developed the important technique of "spectroscopic parallaxes." Although this technique expanded his data base, Kapteyn did not find that it altered his earliest results. Consequently, even though Kapteyn remained sensitive to the validity of the dispersion of the measured parallaxes about their mean, he continued to assert the reliability of his earlier results. In 1920 he summed up his views on this question when he wrote, in his classic paper that formed the basis to his so-called Kapteyn Universe: "It has been shown in G.P. 11 [1901] that widely differing assumptions as to the dispersion law lead to results that differ but little.... Therefore, we have not deemed it necessary to derive this law anew, but

---

[26] H. v. Seeliger, "Die Vertheilung der Sterne auf der nordlichen Halbkugel nach der Bonner Durchmusterung," *Sitzungsberichte der Mathematisch-Physikalischen Klasse der K. Bayerischen Akademie der Wissenschaften zu München*, xiv (1884), 521–48 (here after *München Ak. Sber.*); and idem, "Ueber die Vertheilung der Sterne auf der sudlichen Halbkugel nach Schönfeld's Durchmusterung," *München Ak. Sber.*, xvi (1886), 220–51.

[27] A dependence relationship between density and galactic longitude was not rigorously confirmed until 1917, and so it had little effect on these early developments; see H. Nort, "The Harvard Map of the Sky and the Milky Way," *Recherches astronomiques*, vii (1917), 1–117. After 1918, with the emergence of Shapley's cosmology, and its increasing acceptance by astronomers after 1920, the question of the relation of density distribution to longitude no longer remained of immediate importance; see Chapter 8.

have adopted the one found and tabulated in G.P. 8 [1901]."[28] This was an important point, because the dispersion determines the parameters of the mean-parallax formula.

RESEARCH PROBLEMS AND STELLAR DISTRIBUTIONS, CIRCA 1910

Although these problems dominated research in statistical astronomy to about 1910, they mostly dealt with the large-scale macro-features of the Universe, such as whether or not star-light is absorbed in interstellar space, the position of the Sun with regard to galactic longitude and latitude, and the general distribution of stellar luminosity, stellar spectra, and stellar parallaxes. Through the nineteenth century and into the early years of the twentieth, the primary source of empirical data needed for the study of these problems centered on the counts and motions of stars and their stellar brightnesses. Almost without exception it was universally held that, as the Dutch astronomer Antonie Pannekoek (1873–1960) expressed it in 1910, "all researches into the structure of the universe must start from the knowledge of the function [of] the number of stars of a given magnitude."[29]

After about 1910, however, when the initial phase that defined this research tradition ended, the basic concerns, as understood by Kapteyn, Seeliger, and those doing mainline research into statistical *cosmology,* had become considerably expanded and more sophisticated. Basically they began to focus on two distinct though related classes of problems: first, those problems that explore directly the interrelationship between the most fundamental stellar phenomena, particularly (1) between spectral type and stellar distribution, (2) between mean-parallax, proper motions, and apparent magnitudes, and (3) between stellar distribution and galactic latitude and longitude; and second, those problems that require considerable mathematico-statistical grounding, especially (4) the

[28] J. C. Kapteyn and P. van Rhijn, "On the Distribution of the Stars in Space Especially in the High Galactic Latitudes," *Mount Wilson Observatory, Contributions,* no. 188 (1920), reprinted in *Astrophysical Journal,* 52 (1920), 29 (23–38).

[29] A. Pannekoek, "Researches into the Structure of the Galaxy," *Royal Academy of Amsterdam, Proceedings,* 13 (1910), 240 (239–58).

mathematical form of the luminosity function, and the absolute magnitude at which it obtains a maximum, (5) the analytic form of the density law, (6) the analytical form of the star-count function, including its maximum magnitude value, and the nature of the velocity law and star-streaming in general.

Fundamental stellar phenomena

By about 1914 it was widely believed that the various research problems and their relationship to the larger program of understanding the form and structure of the stellar system had been clearly formulated. To be sure, it was not understood how, for instance, the variation of stellar distribution either along the galactic longitude or with stellar spectra would modify the final conceptions of the sidereal universe. But it was generally assumed that these issues as well as other problems entailed at most a refinement of the various numerical values of the relevant parameters. This is not to suggest, however, that subsequent stellar models would not differ in substantive ways from their earlier formulations. It was simply that the basic techniques had already been formulated so that the newer models included a working out of existing conceptions.

Spectral studies research became unusually active following the discoveries of Hertzsprung and Russell, who showed that there exists a fundamental correlation between absolute magnitude and the stellar spectrum. Although Hertzsprung and Russell independently developed this relationship between 1906 and 1913, it was not until 1914, when Russell made famous their discovery in his classic paper "Relations between the Spectra and Other Characteristics of the Stars," that it became a generally acknowledged empirical fact. Many astronomers remained skeptical of the enormous range in stellar dimensions that the correlation revealed. But with the development of spectroscopic parallaxes by Adams in 1914 and the first successful measurement of the angular diameter of a star predicted by Eddington and confirmed at Mount Wilson in 1920 the mass–luminosity relationship of Hertzsprung and Russell came into general acceptance.[30]

[30] For a thorough analysis of the development of the H–R diagram, see DeVorkin, "Stellar Evolution and the Origin of the Hertzsprung–Russell

Initially Seeliger, Kapteyn, Eddington, Pickering, and Charlier in particular all wrote on the problem of stellar spectra from a mathematical point of view. Because of their mathematical approach, however, spectra investigations introduced overwhelming complications into research on the sidereal problem in general and the density function in particular. The first sustained research into how stellar spectra relates to the distribution of the stars originated with Pickering in 1909. It was only after 1925, however, when strict mathematical solutions were no longer sought, that significant progress was made. As Kapteyn argued in 1918:

> The Bo stars ... [are] no less than 10 million times more luminous than the M stars. In such circumstances the mixing up of all the spectral classes must singularly diminish the effectiveness of any statistical treatment. It is as if we investigated statistically the size of all the members of the animal kingdom from the biggest to the smallest. It must be evident how much more effective must be the treatment of smaller groups such as the genera or species.[31]

Traditional areas of research, such as the determination of mean-parallaxes and the mean distances to various groupings of stars, continued to receive modest but steady support. Because the derivation of the luminosity law, whether using Kapteyn's numerical method (1901) or Schwarzschild's analytical approach (1910),[32] required knowledge of the mean-parallax distribution, studies of this relationship were crucial. "In the investigations about the distribution of the stars in space," wrote van Rhijn in 1915, "the [mean-parallax] function ... has played an important part and it is of great moment to know this function as accurately

Diagram," pp. 90–108, David H. DeVorkin, "The Origins of the Hertzsprung–Russell Diagram," in A. G. Davis and David H. DeVorkin, eds., *In Memory of Henry Norris Russell* (Dudley Observatory Report no. 13, 1977), pp. 61–77, and A. V. Nielsen, "Contributions to the History of the Hertzsprung–Russell Diagram," *Centaurus* ix (1964), 219–53.

[31] J. C. Kapteyn, in J. C. Kapteyn, P. van Rhijn, and H. A. Weersma, "The Secular Parallax of the Stars of Different Magnitude, Galactic Latitude and Spectrum," *Groningen Publications*, no. 29, 1918, p. vii.

[32] For a discussion of Schwarzschild's innovative analytical approach to stellar statistics, see below.

as possible."[33] Most of the investigations undertaken prior to van Rhijn's work of 1923, however, tended to confirm the mean-parallax formula originally suggested by Kapteyn in 1901.[34] It was only after 1920, when large numbers of Mt. Wilson observations of spectroscopic parallaxes and luminosities had become available, that real progress was made.[35]

From William Herschel in the eighteenth century to Kapteyn and Seeliger in the twentieth, stellar count data had received the greatest attention. Thus, for instance, "among the various classes of data needful for a successful attack upon the great problem of the structure and constitution of the stellar universe, none is more fundamental than the numbers of the stars classified according to their apparent brightness and their position in the sky."[36] All statistical astronomers understood this central dogma of their discipline. This data compelled two fundamental questions: (1) What is the rate of increase of the numbers of stars per magnitude class? and (2) What is the relation of this rate to the galactic latitude? It was thought that answers to these queries would provide the form of the star count function that, in turn, was necessary for the solution to the "fundamental equations." Furthermore, this function would yield a reasonably accurate estimate of the total number of stars in the stellar universe. Together with the density law, the absolute distribution of the stars could then be derived.

[33] P. J. van Rhijn, "Derivation of the Change of Colour with Distance and Apparent Magnitude Together with a New Determination of the Mean Parallaxes of the Stars with Given Magnitude and Proper Motion" (Groningen, unpublished Ph.D. diss., 1915), p. 40.

[34] See P. van Rhijn, "On the Mean Parallax of Stars of Determined Proper Motion, Apparent Magnitude and Galactic Latitude for Each Spectral Class," Groningen Publications, no. 34, 1923, 105 pp.; see R. A. Robb, "The Correlation Between Absolute Magnitude, Linear Tangential Velocity, Distance, Apparent Magnitude and Proper Motion," Royal Astronomical Society, Monthly Notices, xmvii (1936), 67–75, for a discussion of the methods developed after 1920 to determine the mean-parallax relationship.

[35] The first results using the new spectroscopic method of determining parallaxes were published by W. S. Adams and A. H. Joy, "The Luminosities and Parallaxes of Five Hundred Stars – First List," Astrophysical Journal, 46 (1917), 313–39.

[36] S. Chapman and P. J. Melotte, "The Number of Stars of Each Photographic Magnitude down to 17.0 m in Different Galactic Latitudes," Royal Astronomical Society, Memoirs, 60(4) (1914), 145 (145–73).

The basic technical problem of these investigations involved the limiting magnitude of the telescopes: Stars beyond a certain stellar brightness could not be observed and hence not counted. If the *shape* as well as the various parameters of the star-count function could be obtained, then a reasonable extrapolation to the remaining stars might be possible. Among those deeply involved in this issue were the Dutch astronomers van Rhijn and H. Nort, the British astronomers S. Chapman, Philibert J. Melotte, and Herbert H. Turner, the Mount Wilson astronomer Seares, as well as Kapteyn, Eddington, Schwarzschild, and Charlier.[37] Beginning with Kapteyn's 1908 pioneering study, considerable discussion and debate ensued between many of the principal investigators. By 1917 van Rhijn, Chapman, Seares, and Nort had all independently vindicated, though partially qualified, Kapteyn's earlier conclusions of the precise *rate* of increase in which the galactic concentration of stars increases with higher magnitude.[38] Because of the more accurate data, van Rhijn's 1917 researches remained the standard source for the surface distribution of the stars until 1925, when the counts from the *Mt. Wilson Catalogue of Selected Areas* became available.

Mathematico-statistical foundation

Except for Kapteyn, most of the leaders in statistical astronomy at this time played a central role in dealing with problems that received a mathematico-statistical grounding. This is not to suggest that Kapteyn was mathematically naive. But as Eddington

[37] See Kapteyn, "On the Number of Stars of Determined Magnitude and Determined Galactic Latitude," pp. 1–14; F. H. Seares, "Preliminary Note on the Distribution of Stars with Respect to the Galactic Plane," *National Academy Proceedings*, 3 (1917), 218 (217–22); P. van Rhijn, "On the Number of Stars of Each Photographic Magnitude in Different Galactic Latitudes," *Groningen Publications*, no. 27 (1917), 33–4; and S. Chapman, "The Number and Galactic Distribution of the Stars," *Royal Astronomical Society, Monthly Notices*, 78 (1917), 66 (66–77).

[38] Van Rhijn, "On the Number of Stars of Each Photographic Magnitude in Different Galactic Latitudes." For the fainter stars (magnitude 10.0 to 15.5) van Rhijn used sixty-five northern *Durchmusterung* photographic plates from the "Plan of Selected Areas" project proposed by Kapteyn in 1906 and taken at the Harvard Observatory.

correctly noted, "Seeliger, Schwarzschild, Charlier, and the English investigators have generally proceeded analytically by expressing the data empirically as mathematical functions. Kapteyn, on the other hand, works to a large extent numerically, as it were sorting the stars into pigeon-holes."[39] Still, Kapteyn was not adverse to this distinction: He recommended Schwarzschild and Eddington as among the "very best men" to Hale precisely because both were mathematically rigorous in their work in stellar studies.[40] Nevertheless, all of these scientists were highly gifted mathematically, and consequently they applied their not insignificant talents to the mathematization of stellar astronomy.

Karl Schwarzschild (1873–1916) began his university training at Strassburg from which he later went to Munich specifically to study with Seeliger.[41] In later life he noted that Seeliger had made a lasting impression on his own scientific work, particularly on Schwarzschild's investigations in stellar statistics to which Schwarzschild contributed significantly. The intellectual atmosphere at Munich was conducive to the interests and abilities of the younger man. Schwarzschild's greatest scientific talents included his ability to apply a critical mathematical outlook to a wide variety of problems in the physical sciences. Although he possessed a clear command of theoretical physics, optics, and electricity, today he is mostly remembered for his contributions to the mathematics of space curvature and the application of general relativity to cosmology.[42]

---

[39] A. S. Eddington, "The Statistical Laws of Stellar Astronomy," *Observatory*, 41 (1918), 385 (384–6).

[40] J. C. Kapteyn to G. E. Hale, 23 September 1915 (Hale).

[41] See O. Bl%menthal, "Karl Schwarzschild," *Jahresbericht der Deutschen Mathematikervereinigung*, 26 (1918), 56–75; A. Einstein, "Karl Schwarzschild," *Sitzungsberichte der Preussischen Akademie der Wissenschaften zu Berlin*, 1916 (1916), 768–70; and S. Oppenheim, "Karl Schwarzschild – Zur 50. Wiederkehr seiner Geburtstages," *Astronomische Gesellschaft Vierteljahrschrift*, 58 (1923), 191–209. Schwarzschild's brilliant scientific career was cut short at age 43 when he developed symptoms of a rare, painful, and at the time incurable illness (pemphigus) while serving on the Russian front during the First World War.

[42] Sally H. Dieke, "Karl Schwarzschild," in C. C. Gillispie, ed., *Dictionary of Scientific Biography* (New York: Charles Scribner's & Sons, 1975, 16 vols.), vol. 12, pp. 247–53.

Figure 13   Karl Schwarzschild, ca. 1910. (Courtesy of Yerkes Observatory)

Seeliger acknowledged in 1909 that he did not possess a general analytic solution to the "fundamental equation." Because its solution was thought necessary to represent a comprehensive understanding of the (theoretical) distribution of the stars, the necessity of a reliable solution became a primary task for statistical cosmologists. Within a year Schwarzschild provided the very solution that had previously escaped Seeliger's analysis.[43] Briefly, assuming empirical forms for the star-count and mean-parallax functions, Schwarzschild was able simultaneously to solve the "fundamental equation of stellar statistics" and the "fundamental equation for the mean parallax" for the density and luminosity relationships using fourier transformations. The essence of the fourier transform lies in the fact that it can be used to "invert" an integral equation, that is, to transpose the integral function with its integrand. This technique transforms the given equation into an equation solvable with standard tools available in the integral calculus.[44] Schwarzschild realized that the solution to both relationships (density and luminosity) was based on the general mathematical principle that as many unknowns could be solved for as one possessed independent equations. As a result, Schwarzschild raised the mathematics of stellar statistics to a new – and theoretically more rigorous – level by assuming prior knowledge of only the two basic empirical functions (star-counts and mean-parallaxes).

Schwarzschild's approach to the fundamental equations using fourier analysis provided a clear alternative to Kapteyn's (numerical) approach to the determination of the luminosity relationship. Both methods, however, required explicit knowledge of the same empirical data (star-counts and mean-parallaxes). By

[43] K. Schwarzschild, "Ueber die Integralgleichungen der Stellar-Statistik," *Astronomische Nachrichten*, 185 (1910), 81–8; also see R. Kurth, "Zur Schwarzschilden Integralgleichung," *Zeitschrift für Astrophysik*, 31 (1952), 115–20.

[44] Fourier's invention of "transformations" stems from his famous 1811 paper on the theory of heat. Although he did not conceive of his method in such terms, mathematically this work belongs to the history of integral equations. Analytically his developments were equivalent to such a solution and thus, in a sense, he was the first mathematician to provide the solution of an integral equation. See M. Kline, *Mathematical Thought from Ancient to Modern Times* (New York: Oxford University Press, 1972), pp. 1052–3.

providing an analytic solution to Seeliger's formulation of the sidereal problem, Schwarzschild not only further secured the mathematical basis to stellar statistics but he also fundamentally emphasized the prominence of the mean-parallax and star-count functions. Regardless of which method was chosen, the numerical (Kapteyn) or the analytical (Schwarzschild), knowledge of these functions became crucial for an understanding of the luminosity and density relationships.

Seeliger remained skeptical, not of Schwarzschild's innovative approach, but of such crucial reliance on an empirically derived mean-parallax function. Moreover, not only did Schwarzschild rely exclusively on Kapteyn's formulation of this relationship (and the star-count function), but Schwarzschild's density relationship was not fully commensurate with the one that Seeliger believed to represent the actual state of affairs. The problem, Seeliger clearly realized, derived from the assumed nature of the empirical functions. Consequently, in his third major investigation of the distribution of the stars in 1911, Seeliger postulated a rather complex form of the luminosity relationship and a slightly modified version of his earlier (1898) density law. This allowed him to approximate the empirical data of star-counts and mean-parallaxes without assuming their analytic form.[45]

For nearly a century, the only reasonably reliable stellar function was star-counts, representing the numbers of stars within each stellar (apparent) magnitude class. Its significance stemmed not from its historical priority, however, but from its logical relevance to the various fundamental equations. As Kapteyn expressed in 1908: "In the investigation of the arrangement of the stars in space, there [is] no element which has played so prominent a part as the number of stars of a determined magnitude and the variation of this number with the galactic latitude."[46] This view, reiterated by van Rhijn, continued unabated: "An accurate knowledge of the number of stars of determined magnitude is of very great importance in stellar statistical investigations. For these numbers are closely connected with the density and luminosity laws, which

[45] H. von Seeliger, "Ueber die räumliche Verteilung der Sterne im schematischen Sternsystem," *München Ak. Sber.*, 41 (1911), pp. 442–3 (413–61).
[46] Kapteyn, "On the Number of Stars of Determined Magnitude and Determined Galactic Latitude," p. 1.

are two of the principal functions determining the structure of the Milky Way system."[47]

Not only did van Rhijn reemphasize the centrality of star-counts, but, most important, by claiming the necessity of the density and luminosity relationships for an understanding of the stellar system, he reaffirmed the hegemony of statistical astronomy for cosmological investigations during the early years of the present century to about 1920.

As early as 1885, Edward Pickering had already empirically derived a nonanalytic, first-order (linear) approximation for star-counts. Drawing on William Herschel's stellar assumptions, however, Pickering later (1902) admitted its inadequacy.[48] Rather than infer its analyticity from the empirical data directly as Pickering and others had attempted, Kapteyn, using Seeliger's fundamental equation of stellar statistics, derived a star-count function mathematically in 1904,[49] and, in 1908, extended its reliability down to a limiting apparent magnitude of 19.0, a full ten magnitudes beyond that found in the *Bonner Durchmusterung*.[50] Based on Kapteyn's 1908 work, Schwarzschild obtained a second-order approximation for the star-count function that subsequently became the basis for virtually all future investigations.[51]

Actually, the star-count function is simply a *statistical* representation of the numbers of stars corresponding to their apparent magnitude (and galactic latitude). Therefore, it possesses both a "mean value" and a "dispersion" describing the width of the statistical spread of the apparent magnitude or independent variable. Consequently, by its very nature it is a small step to wed the empirical data with statistical theory. That step was most

---

[47] Van Rhijn, "On the Number of Stars of each Photographic Magnitude in Different Galactic Latitudes," p. 1.

[48] E. C. Pickering, "Distribution of Stars," *Harvard Astronomical Observatory, Annals*, 48 (5) (1902–3), 150–3 (149–85); also see E. C. Pickering, "Distribution of the Stars," *Harvard Astronomical Observatory, Circular*, 147 (1909), 1 (1–4).

[49] J. C. Kapteyn, "Remarks on the Determination of the Number and Mean Parallax of Stars of Different Magnitude and the Absorption of Light in Space," *Astrophysical Journal*, 24 (1904), 115–17 (115–23).

[50] Kapteyn, "On the Number of Stars of Determined Magnitude and Determined Galactic Latitude," 18 (1908).

[51] K. Schwarzschild, "Ueber die Integralgleichungen der Stellar-Statistik," *Astronomische Nachrichten*, 185 (1910), 81–8.

forcefully taken first by the Swedish mathematical astronomer Carl Vilhelm Ludwig Charlier (1862–1934), who after 1910 became one of the strongest advocates for a full mathematization of stellar statistics.[52] His influence, particularly in Sweden, was extensive; among those who felt his guiding hand were Bertil Lindblad and Karl G. Malmquist. From his earliest years at the University of Upsala, Charlier, as with Seeliger and Schwarzschild, seemed most comfortable with things mathematical and theoretical. Indeed, because of Charlier's theoretical propensity in mathematics and statistics, he always pursued foundational issues to the very bottom by exploring rigorously the most general mathematico-statistical formulations. Shortly after obtaining his Ph.D. in 1887, Charlier became an assistant to Hugo Gyldén at the Stockholm Observatory. Gyldén, who had been one of the two most prominent celestial mechanists of the nineteenth century and a close personal friend of Seeliger's, directed Charlier into planetary physics, where he first applied his mathematical talents to celestial mechanics.[53] Within a decade he was appointed professor of astronomy and director of the observatory at Lund, where he remained until his retirement in 1927.

Charlier's interests in applying a statistical methodology to astronomy has a long and distinguished history. As recognized early in the nineteenth century by the Belgian statistician Adolphe Quetelet (1796–1874), the astronomer's love of natural order provided the foundation for statistical science:

> It is not to doctors that we owe the first tables of mortality, they were calculated by the celebrated astronomer [Edmond] Halley. . . . The laws that concern man, and those that govern social development, have always had a special attraction for the philosopher, and perhaps most especially for those who have directed their attention to the system of the universe.[54]

[52]   R. J. Trumpler, "Address of the Retiring President of the Society in Awarding the Bruce Medal to Prof. C. V. L. Charlier," *Astronomical Society of the Pacific, Publications*, 45 (1933), 5–14.

[53]   See correspondence of H. v. Seeliger to H. Gyldén, 1889 to 1896 (Stockholm).

[54]   Quoted in Theodore M. Porter, *The Rise of Statistical Thinking, 1820–1900* (Princeton: Princeton University Press, 1986), p. 44.

Figure 14   Carl Vilhelm Ludwig Charlier, ca. 1925.

Charlier became convinced that the mass of stellar data neces-sitated a rigorous statistical foundation.[55] As a consequence, it was during his early years at Lund that Charlier undertook a critical examination of the investigations of the English statisti-cian Karl Pearson, thus becoming one of the most prominent members of the so-called Scandinavian School of Statisticians.[56] Initially his studies dealt only with theory; later he began to make applications to a variety of fields, including biology, medicine, population statistics, as well as stellar astronomy, of course. At Lund he eventually inaugurated courses in statistics, an academic program that, after his retirement, led to the establishment of a sep-arate chair for mathematical statistics. Although Seeliger, Kapteyn, and to a lesser extent Schwarzschild all had applied various statistical concepts and methods to a variety of astronomical problems, Charlier's investigations of the mathematical founda-tions of statistics represented altogether a different dimension to the stellar statistics that the others were pursuing. Being in some respects first a statistician, Charlier added a rigorous *statistical* foundation to stellar statistics unheard of hitherto.

Charlier, who was deeply versed in mathematics and statistics, turned his not insignificant mathematical prowess first to the foundation of mathematical statistics and then, around 1912, to statistical cosmology.[57] By 1912 the work of Seeliger and Kapteyn

---

[55] Beginning about 1905 Charlier published a long series of papers on the foundations of probability theory and the mathematical theory of statistics. See Charlier's papers in *Meddelanden Fran Lunds Astronomiska Obser-vatorium* (Serie I), nos. 25, 26, 27, 34, 41, 49, 50, 51, 52, 57, 58, 61, 62, 71 (hereafter *Lunds Medd*).

[56] H. M. Walker, *Studies in the History of Statistical Method* (Baltimore, 1931), pp. 81, 163. The most prominent members of this school were L. H. F. Opperman, J. P. Gram, T. N. Thiele, H. Westergaard, and Charlier. For an examination of the development of mathematical statistics in astronomy, see Stephen M. Stigler, *The History of Statistics: The Measurement of Uncer-tainty before 1900* (Cambridge, Mass.; Harvard University Press, 1986), pp. 11–158, Jerzy Neyman, "The Emergence of Mathematical Statistics: A His-torical Sketch with Particular Reference to the United States," in D. B. Owen, ed., *On the History of Statistics and Probability* (New York: Marcel Dekker, Inc., 1976), pp. 149–93, and Elizabeth L. Scott, "Statistics in As-tronomy in the United States," in Owen, *On the History of Statistics and Probability*, pp. 319–31.

[57] C. V. L. Charlier, "Studies in Stellar Statistics – Constitution of the Milky Way," *Lunds Medd* (II), 8 (1912) (63 pp.).

had securely fixed the study of the distribution of the stars to its mathematical and statistical foundation. As with the other prominent researchers, Charlier noted that the two principal goals of statistical cosmology were the determination of the space distributions of stars (density and luminosity relationships) and the description of stellar motions (velocity relationship). More gifted mathematically than either Seeliger or Kapteyn, Charlier took the results of his older colleagues and provided a more rigorous mathematical foundation to the study of statistical cosmology.[58] Thus beginning in 1912 Charlier produced a series of studies on these research problems that relied fundamentally on his previously rigorous background in statistics. Convinced of Seeliger's general approach, he rejected Schwarzschild's (and Kapteyn's) use of an empirically inferred mean-parallax function, because "in practice it is found difficult to determine the mean distances of the stars in such numbers that the form of the [mean-parallax] function ... can be considered as known with a sufficient degree of approximation."[59] Charlier adopted Schwarzschild's expression of the star-count function, however, expressing it rigorously in statistical terms. Moreover, he argued that the same "normal" form could be used for the luminosity relationship, something that he realized Kapteyn had already concluded without knowing the reason.

As a rigorous statistician, Charlier was particularly capable of explaining why the *form* of a mathematical expression is equally important to its content. In order to demonstrate the superiority of his more rigorous theory, Charlier exposed the inadequacies of Schwarzschild's 1910 study (which in turn was derived from Kapteyn's investigations) by using Schwarzschild's mean value of 51.0 for the apparent magnitude of all stars in the Galaxy. This corresponded to a total number of stars in the stellar universe of about 50 trillion, a value nearly 1 million times greater than stellar estimates accepted at the time (and about 100 times greater than contemporary estimates for the entire Galaxy). As Charlier deliberately underscored, this value is "in direct contradiction to

---

[58] C. V. L. Charlier, "Contributions to the Mathematical Theory of Statistics," *Lunds Medd* series I (1906–12) and series II (1912–16).
[59] Charlier, "Studies in Stellar Statistics – Constitution of the Milky Way," *Lunds Medd* (II), p. 6.

the usual view of our starry system."[60] Charlier's main purpose, however, was not to criticize the reliability of Schwarzschild's or Kapteyn's theories. Rather it was a dramatic way to demonstrate the importance of incorporating statistical concepts and methods into the mathematical theory of stellar statistics.

Because the verification of the luminosity relationship was based on mean-parallaxes, the luminosity relationship was valid only within the local solar neighborhood. Given this restriction, stellar astronomers further assumed, under the supposition of negligible light absorption, that the statistical spread of the luminosity-curve must be nearly uniform throughout space. Until at least the mid-1920s, these assumptions as well as the analytical form of the luminosity relationship remained basic results in stellar statistics. From this time on (1912), therefore, Charlier's representations for the star-count function and the luminosity relationship became standard in the formal discussions of stellar statistics.

RESEARCH PROBLEMS AND STELLAR MOTIONS, CIRCA 1910

The last general problem area, that dealing with the nature of the velocity law and star-streaming in general, has a development somewhat independent of – though certainly related to – the other major problems. As we have already noted in Chapter 4, Kapteyn's discovery of star-streaming in 1902 eventually stimulated considerable interest in "preferential" stellar motions. It was not Kapteyn's 1904 announcement of his discovery that elicited the large response, however, but his announcement the following year at the annual meetings of the British Association for the Advancement of Science.[61] Not only did the 1905 paper focus exclusively on star-streaming, but the BAAS was more directly in the mainstream of scientific activity than St. Louis.[62]

---

[60] Ibid., pp. 21–2.

[61] J. C. Kapteyn, "Statistical Methods of Stellar Astronomy," *International Congress of Arts and Sciences* (St. Louis, 1904), vol. 4, pp. 412–22, and J. C. Kapteyn, "Star-streaming," *British Association for the Advancement of Science, Report*, section A (1905), 257–65.

[62] See Katherine R. Sopka, "Preface," and Albert E. Moyer, "Foreword," in Katherine R. Sopka, ed. and comp., *Physics for a New Century: Papers Presented at the 1904 St. Louis Congress* (Los Angeles: American Institute of Physics, 1986), pp. ix–xx.

More so than any other investigator – including Kapteyn – Arthur S. Eddington (1882–1944) produced so many fundamental research papers on star-streaming that his name became virtually synonymous with this phenomenon.[63] Indeed, in the words of the English astronomer-historian J. L. E. Dreyer, Eddington's mathematical theory of star-streaming raised the phenomenon of star-streams "to the rank of an important discovery."[64] In his first paper in 1906, Eddington outlined the mathematical principles of star-streaming that dominated his thinking on the issue. In his analysis, he argued that for "the best mathematical equivalent of 'haphazard' [motions], a distribution of velocities according to Maxwell's law" should prevail.[65] Suggested first in 1859 by the Scottish physicist James Clerk Maxwell, Maxwell's law became the first major natural law that is statistical in nature, and not absolute, deterministic, or causal.[66] Eddington was hoping to extend this general statistical law, called "one of the most beautiful and fruitful of all applications of probability theory to physics," into astronomy.[67] Although Maxwell's law is connected with the kinetic theory of gases, Eddington argued (relying on an argument by the English astronomer-physicist James Jeans) that a Maxwellian distribution is "the most probable."[68] Consequently, Eddington suggested that a distribution of stellar velocities representing the two star-streams could be accounted for by the superposition of two separate Maxwellian distributions.

Neither Kapteyn's explanation nor Eddington's hypothesis, which came to be known as the "two-drift theory," were the only

[63] A. V. Douglas, *The Life of Arthur Stanley Eddington* (London, 1957), p. 20.
[64] J. L. E. Dreyer, as quoted in ibid.
[65] A. S. Eddington, "The Systematic Motions of the Stars," *Royal Astronomical Society, Monthly Notices*, 67 (1906), 35 (34–63). Eddington published extensively on the phenomenon of star-streaming; see esp. ibid., and Eddington's masterly summary of the issue in his *Stellar Movements and the Structure of the Universe* (London: Macmillan & Co., 1914). For the background to Eddington's efforts to formulate a mathematical theory of star-streaming, see Douglas, *Eddington*, pp. 16–17, 20–3.
[66] J. C. Maxwell, *Collected Works*, vol. 1, p. 380; also see Edna E. Kramer, *The Nature and Growth of Modern Mathematics* (New York: Hawthorn Books, Inc., 1970), pp. 345–6.
[67] Emile Borel, quoted in ibid., p. 346.
[68] See J. Jeans, *The Dynamical Theory of Gases* (1904), section 56, and Eddington, "The Systematic Motions of the Stars," p. 35.

Figure 15   Arthur S. Eddington, ca. 1920. (Courtesy of Yerkes Observatory)

models of star-streaming. In 1907 Schwarzschild rejected Eddington's theory on the grounds that it implied the existence of a definite physical difference between the stars of the two streams.[69] If the stars of the two-drift theory actually consist of two intermingled subpopulations, it would not be unreasonable to expect other differences among the stars of the two drifts, such as differences in brightness, spectral type, and density, in addition to differences in their motions. Because no such nonkinematic features were observed, on conceptual grounds Schwarzschild rejected Eddington's theory in favor of the more fundamental principle that the Universe, though manifesting a wide variety of phenomena, is nevertheless to be understood in strictly uniformitarian and parsimonious terms. Mathematically, Schwarzschild showed that star-streaming could be interpreted not only as two independent star streams, but also as a *unitary* system whose velocity distribution could be represented by an ellipsoidal expression in which each axis corresponded to one of the two dominant star-streams. Even though there are three velocity components in three space, Schwarzschild combined the two minor components representing the resulting distributions using Maxwell's law.

Probably because Eddington and Schwarzschild both recognized one another's considerable mathematical prowess, their correspondence is remarkably genteel and noncombative. Still, Eddington recognized an empirically sensitive, kinematic difference between their two theories. Within a few months of publication of the 1907 paper, Eddington wrote Schwarzschild about their "laws":

> The main difference between the two laws must be that mine leads to more of a neck [subsequently called "Eddington's rabbits"], and I cannot help thinking that the observations tend to support my law – though that is merely a matter of opinion. Again although your theory is mathematically simpler than mine, it appears to me much less simple physically.

[69] K. Schwarzschild, "Ueber die Eigenbewegung der Fixsterne," *Nachrichten von der K. Gesellschaft der Wissenschaften zu Göttingen* (1907), 615 (614–32). Also see R. J. Trumpler and H. F. Weaver, *Statistical Astronomy* (Berkeley: University of California Press, 1953), p. 278.

Eddington quickly reassured Schwarzschild that "although your point of view is somewhat different from what mine has been, our hypotheses seem to me to lead to a very closely similar law of distribution of the proper motions in space."[70] Following publication of Schwarzschild's (February) 1908 paper on refining the mathematics of his ellipsoidal hypothesis, Eddington responded:

> You will see that we have both been attacking the same problem – how to deal with sparsely distributed motions; but I must confess that my solution, although it was to me almost unexpectedly successful, seems rather a failure now that it has to be compared with your delightfully simple method.[71]

Schwarzschild and Eddington's contemporary Charlier championed the view that the stellar motions are not caused by the intermingling of two star-streams moving through one another (Kapteyn), but rather may be explained using (Schwarzschild's) ellipsoidal distributions.[72] At the Hamburg meetings of the Astronomische Gesellschaft in 1913, Charlier illustrated his view by comparing Kapteyn's two-stream hypothesis with Ptolemy's epicyclic theory of planetary motions, whereas Schwarzschild's ellipsoidal hypothesis corresponded analogously to Kepler's theory of elliptical orbits. After Schwarzschild's death in 1916, Charlier became the premier advocate of the ellipsoidal hypothesis, giving it its strongest mathematical foundation.[73]

Although intense controversy arose concerning the relative merits of the two explanations, Eddington, Schwarzschild, and Charlier did not participate in the debate. For example, Eddington's compatriot Jacob Halm, writing to Schwarzschild in 1910, expressed the view that

[70] A. S. Eddington to K. Schwarzschild, 20 January 1908 (Schwarzschild).
[71] A. S. Eddington to K. Schwarzschild, 11 September 1908 (Schwarzschild).
[72] Charlier, "Studies in Stellar Statistics – The Motion of the Stars," *Lunds Medd* (II), no. 9 (1912); Charlier, "Do the Star Streams of Kapteyn exist?" *Astronomical Society of the Pacific, Publications*, 36 (1924), 212–15.
[73] C. V. L. Charlier, *The Motion and Distribution of the Stars* (Berkeley, 1926), pp. 3–82. This monograph is based on a series of lectures given by Charlier at Berkeley in 1924 that provided a technical exposition for research astronomers similar to Eddington's earlier book *Stellar Movement and the Structure of the Universe* (1914).

it would seem that the two drift hypothesis [Eddington] explains the phenomena [sic] of the Bradley stars better than the ellipsoidal theory, but I am not sure that a modification of your brilliant analysis by introducing the assumption of an eccentric position of our system [the Sun] with reference to your velocity ellipse might not account for the facts referred to in our papers.[74]

In collaborative work, Eddington and Halm went on to argue for the existence of a third stellar "drift,"[75] which Eddington (and Schwarzschild) later recognized as a minor fluctuation of the basic phenomenon of star-streaming.[76]

Although the physical implications of Eddington's and Schwarzschild's theories were radically different, in fact both Schwarzschild and Eddington expressed the view that except for the mathematical form the two theories are indistinguishable.[77] "Although [Schwarzschild's] theory contemplates the universe as single, instead of as composed of two distinct systems," wrote Eddington in 1910, "there is really but little difference between the actual distributions of the motions which the two theories formulate. They express nearly the same law, but by the aid of different mathematical functions."[78] Prior to the recognition of "differential galactic rotation" by Bertil Lindblad and Jan Oort during the later 1920s, the explanatory power of both Eddington's and Schwarzschild's theories was nearly identical.[79]

Before the discovery of star-streaming, Kapteyn (as we have already seen in Chapter 4) had utterly failed to derive both the velocity and density laws because his approach was based almost

---

[74] J. Halm to K. Schwarzschild, 14 September 1910 (Schwarzschild).
[75] A. S. Eddington, "The Systematic Motions of the Stars of Prof. Boss's 'Preliminary General Catalogue'," Royal Astronomical Society, Monthly Notices, 71 (1910), 4–42. The "third stellar stream" theory was further developed by J. Halm in his "Further Considerations relating to the Systematic Motions of the Stars," Royal Astronomical Society, Monthly Notices, 71 (1911), 610–39.
[76] A. S. Eddington to K. Schwarzschild, 29 December 1911 (Schwarzschild).
[77] Schwarzschild, "Ueber die Eigenbewegung der Fixsterne," pp. 630–1.
[78] A. S. Eddington, "Star-Streams," Scientia, 8 (1910), 39 (30–40).
[79] Eddington's classic Stellar Movements and the Structure of the Universe also presented equally the advantages of both theories. For a discussion of the relationship of the velocity ellipsoid to the structure and dynamics of the stellar system, see Trumpler and Weaver, Statistical Astronomy, pp. 278–9.

exclusively on stellar motions. It was only after theorists had come to an understanding of preferential motions as reflected in star-streaming that Kapteyn's earlier approach for solving the sidereal problem was renewed. Specifically, it was thought that focusing on stellar motion data would yield a velocity law describing *galactic* motions. In turn this law would be used to derive the density (and luminosity) relationship that would then be used to describe the architecture of the stellar system.

This thinking continued among some specialists until 1912 when Eddington was able to develop such an approach for the derivation of the density relationship. Rather than adopting stellar counts, mean-parallaxes, a luminosity function, and proceeding in the standard manner using the fundamental equations, he suggested an approach somewhat reminiscent of the one Kapteyn had initially employed during the 1890s, before his discovery of star-streaming.[80] Mathematically, Eddington derived an analogue to the fundamental equation of stellar statistics, what we might call the "fundamental equation of stellar motions," from which he obtained the velocity law and, as a by-product, a numerical representation of the density function. His source of ideas for developing this technique came from none other than his friend Karl Schwarzschild. Writing Schwarzschild in September 1911, Eddington revealed:

> I have spent a lot of time working on the lines of your paper on "Integral Equations of Stellar Statistics," only applying the same method to proper motions. Unfortunately, I have not managed to get much out of it as yet; . . . but I hope to get something out ultimately. I was very delighted at finding that there was such an elegant method of solving these equations; it should be most useful in further work on stellar statistics.[81]

Even though Eddington did not derive the density relationship analytically from the velocity relationship in his 1912 work, both he and Frank W. Dyson, Britain's Astronomer Royal at the time, independently assumed a mathematical form for the density

[80] A. S. Eddington, "A Determination of the Frequency-law of Stellar Motions," *Royal Astronomical Society, Monthly Notices*, 72 (1912), 368–86.
[81] A. S. Eddington to K. Schwarzschild, 14 September 1911 (Schwarzschild).

function that allowed them to approximate the observations and empirical data for stellar motions with reasonable precision.[82] Patterned on Schwarzschild's fourier solution of integral equations, the essential data Eddington and Dyson required were based solely on proper motions.[83] Moreover, their results gave the actual numbers of stars, not simply a proportional distribution. This compelled Dyson to write with confidence: "If the law of distribution of the linear velocities of the stars in any part of the sky were known, it should be possible . . . to derive the law of distribution of stars [i.e., density law] according to distance."[84]

Because this approach leads theoretically to a solution of the velocity relationship and because it is empirically independent of the Seeliger–Schwarzschild method for the derivation of the density relationship, "the one," argued Eddington, "is a complete check on the other."[85] Consequently, by about 1915 it was thought that there existed two, largely independent methods of understanding the architecture of the Milky Way system – the first based primarily on magnitude counts and mean-parallactic motions, and the second grounded on the distribution of proper motions.

The mathematical contributions that Eddington and Dyson made consisted in the formulation of the principal laws of statistical astronomy from data based largely on stellar motions. Indeed, both men possessed considerable mathematical prowess. Eddington had been Senior Wrangler or the top mathematics student at Cambridge during his second year, an unprecedented achievement; Dyson achieved second Wrangler, not in itself a mean accomplishment because only two such prizes are awarded annually. (By

---

[82] Eddington suggested the use of Seeliger's 1898 density law; see A. S. Eddington to K. Schwarzschild, 29 December 1911 (Schwarzschild).

[83] A. S. Eddington, "The Distribution in Space of the Bright Stars," *Royal Astronomical Society, Monthly Notices*, 73 (1913), 346–58, and F. W. Dyson, "The Distribution in Space of the Stars in Carrington's Circumpolar Catalogue," *Royal Astronomical Society, Monthly Notices*, 73 (1913), 334–5, 402–6, and F. W. Dyson, "The Proper Motions of the Stars in Carrington's Circumpolar Catalogue in Relation to their Spectral Types," *Royal Astronomical Society, Monthly Notices*, 74 (1914), 733–44.

[84] Dyson, "The Distribution in Space of the Stars in Carrington's Circumpolar Catalogue," p. 334.

[85] Eddington, *Stellar Movements*, p. 230; his method is summarized on pp. 218–29.

comparison, both Lord Kelvin and J. C. Maxwell were second Wranglers in their student days.) As Eddington admitted, though these investigations "suffer from scantiness of data, insufficient range, and the absence of an effective check on the main assumptions, . . . yet I do not think it can be doubted that these general statistical researches have already greatly advanced our knowledge of the distribution and luminosities of the stars."[86] The weaknesses of their approach, which Eddington himself alluded to above, however, turned out to be formidable. Besides a paucity of proper motion data and a limited magnitude range, independent verification of both the velocity and density relationships, neither of which could be empirically confirmed *directly*, was the crucial stumbling block.[87] Consequently, though there were several relatively unproductive applications of their method by a few others, subsequent development of this method was not forthcoming. Even standard textbooks, which present a complete discussion of the first method (developed by Seeliger, Schwarzschild, Kapteyn, and the others), without exception ignore virtually any mention of the second approach.[88]

Although Kapteyn himself eventually rejected the second approach, he remained keenly aware of the theoretical explanations of star-streaming offered by Schwarzschild, as well as by Eddington and Charlier.[89] Results made available by these researchers so

[86] Eddington, *Stellar Movements*, p. 230.
[87] It was not until after 1920 that proper motion data became available for the telescopic stars in both hemispheres.
[88] See R. J. Pocock, "The Distribution in Space of the Stars in Zone +25° of the Oxford Astrographic Catalogue," *Royal Astronomical Society, Monthly Notices*, 76 (1916), 421–8, and H. W. Unthank, "A Determination of the Mean Parallax of Stars for Different Magnitudes," *Royal Astronomical Society, Monthly Notices*, 76 (1916), 529–38. Unthank was a student of Eddington's. At least one stellar investigator, writing as late as 1928, considered Eddington's and Dyson's work a "good first approximation"; see W. H. Smart, "On the Law of Stellar Distribution Derived from Proper Motions," *Royal Astronomical Society, Monthly Notices*, 88 (1928), 567–83. Neither Trumpler and Weaver's *Statistical Astronomy* nor E. von der Pahlen's *Lehrbuch der Stellarstatistik* (Leipzig: Johann Ambrosius Barth Verlag, 1937) mention either Eddington or Dyson in this context. The fundamental equation of stellar motions is presented, however, in R. Kurth, *Introduction to Stellar Statistics* (Oxford: Pergamon Press, 1967), pp. 149–50.
[89] Kapteyn first became aware of Eddington's work from Gill; see D. Gill to J. C. Kapteyn, 10 November 1906 (K.A.L.).

stimulated Kapteyn, however, that in 1912 he returned briefly to a discussion of his star-streaming discovery. Kapteyn's thinking on the utility of these two methods, however, was far from conclusive. By at least 1910 Kapteyn had commenced a warm and fruitful correspondence with Schwarzschild, which continued until the latter's death in 1916. In one of his most revealing private assessments he wrote Schwarzschild in 1912:

> Weersma has used your [density] formula [i.e., the first approach] to calculate the stellar distribution for the higher galactic latitudes for me. . . . I have not been able to examine the results with sufficient care, but it appears to me that they are not conclusive. Naturally, for small distance one must find the same stellar density as with the lower galactic latitudes, or as in the treatment of the entire stellar system. There is a difference, but not nearly as great as I had thought. By Eddington's work [the second method] it is difficult to see which direction his results should apply. His work has encouraged me to publish my own solution, and because I cannot quite get it finished, I have gotten Weersma to get it finished for the press. This he has done and I will send it off to the *Monthly Notices*. My solution also calls out for a two-part system.[90]

Except for his 1912 paper to the *Monthly Notices* in which he examined the empirical foundations as well as the mathematical principles of star-streaming, Kapteyn participated relatively little in the later mathematical developments resulting from his 1902 discovery.[91]

Throughout his scientific career, Kapteyn urged astronomical colleagues worldwide to continue research both of parallactic and of proper motion and radial velocity phenomena. The former data were needed to determine the luminosity and density functions; the latter were required to verify precisely the velocity law and star-streaming. Of his own work on star-streaming he

---

[90] J. C. Kapteyn to K. Schwarzschild, 9 May 1912 (Schwarzschild).
[91] J. C. Kapteyn and H. A. Weersma, "On the Derivation of the Constants for the Two Star Streams," *Royal Astronomical Society, Monthly Notices*, 72 (1912), 743–56.

wrote to Schwarzschild in 1911: "I have spent considerable time on the Orion and early-A stars. Now many radial velocities are being determined at Mount Wilson that are especially important for this investigation. Before I have these [however] it is best to publish nothing."[92] Of Schwarzschild's investigations, he wrote the following year: "You cannot know how pleased I am that you are continuing with the radial velocity work."[93] From these and other investigations, Kapteyn hoped for a thorough explanation of star-streaming, particularly in light of his more central research on the luminosity and density functions. Indeed, influenced deeply by these investigators, Kapteyn, in his last major attempt at modeling the sidereal system (in 1922), explained star-streaming in terms of a gravitational attraction about a rotating system of stars.[94]

METHODOLOGY, STELLAR STATISTICS, AND SCIENTIFIC EXPLANATION

Although Kapteyn's paper of 1901 had emphasized a numerical approach, it did not appear as a clear alternative, nor did it gain the initiative over the elegant analytical methods until after 1920.[95] It was only when strictly analytic approaches became increasingly problematic, due mainly to the irregular density distribution of stars, that the numerical methods came into prominence. Prior to this gradual shift in the methodological emphasis of statistical astronomy, however, most astronomers looked forward to the time when stellar statistics would become a branch of mathematical analysis.

The emphasis on mathematizing stellar statistics was neither at variance with general scientific developments of the period nor was it exclusively an expression of preference influenced by mathematical astronomers such as Seeliger, Schwarzschild,

[92] J. C. Kapteyn to K. Schwarzschild, 12 November 1911 (Schwarzschild).
[93] J. C. Kapteyn to K. Schwarzschild, 9 May 1912 (Schwarzschild).
[94] For a treatment of this issue, see the discussion in Chapter 6 on Kapteyn's 1922 theory of the stellar system.
[95] See Bart Bok, *The Distribution of the Stars in Space* (Chicago: University of Chicago Press, 1937), p. 26.

Charlier, Eddington, and Dyson.[96] To be sure, Seeliger, as well as other mathematical astronomers, was inclined toward a mathematical approach in astronomical methodology. But even this mathematical preference was a reflection of the dominating trend in physical science toward positivistic explanations of natural phenomena in terms of mathematical equations rather than mechanistic explanation. Kapteyn, too, reflected this trend, even though he maintained a belief in the efficacy of the traditional numerical approach to the explanation of stellar phenomena. Like Seeliger, he had been trained as a mathematical physicist and approached stellar studies, particularly during the 1890s and early 1900s, to a large extent within a rigorous mathematical framework. Although Kapteyn continued to use significant mathematics throughout his work, it was only after his mathematical theory of star-streaming failed that he felt particularly disinclined to commit fully to a strict reductionist approach in statistical cosmology.

The trend toward incorporating stellar statistics as a branch of mathematical science so dominated developments in statistical astronomy during the first two decades of the twentieth century that it deserves additional attention. The emphasis on the primacy of mathematical explanation within a positivistic background had been gaining momentum over the last quarter of the nineteenth century. As a methodological tool, mathematics had been increasing in significance since the seventeenth century. Achievements of the eighteenth-century continental mechanists and the French school of mathematical physicists during the Napoleonic era had further enhanced the possibilities of a mathematical approach.[97] In addition to a mathematical procedure, the search for mechanical explanation of natural phenomena was the accepted and largely successful program of physical theorists until the latter part of the nineteenth century.[98] In this search mathematics was used primarily as a tool to provide insight into the actual mechanisms

[96] For an explicit discussion of this trend, see, for example, Eddington, "The Statistical Laws of Stellar Astronomy," p. 385.
[97] See R. H. Silliman, "Fresnel and the Emergence of Physics as a Discipline," *Historical Studies in the Physical Sciences*, 4 (1974), 137–62, and R. Fox, "The Rise and Fall of Laplacian Physics," *Historical Studies in the Physical Sciences*, 44 (1974), 89–136.
[98] M. Kline, "Mechanical Explanation at the End of the 19th Century," *Centaurus*, 17 (1972), 58.

governing the operations of the real world. Even so the various mechanical theories (e.g., the atomic theory, the first law of thermodynamics, the kinetic theory of gases) were not uniformly successful. During the latter decades of the nineteenth century, mechanical explanations only were considered largely inadequate and it was considered that only a mathematical explanation of phenomena, without postulating an underlying materialistic structure, could be counted on to provide a rational explanation of natural events.[99] These developments provided a climate favorable to mathematical modeling within science in general and within astronomy in particular.

The general reaction against mechanistic materialism was in full swing during the last quarter of the nineteenth century. It was based in part on the failure of one of the great projects of nineteenth-century physics: the construction of a mechanical or atomic-molecular model of matter and of the ether from which thermodynamic and electrodynamic phenomena could be explained strictly by means of the laws of mechanics. The developments in thermodynamics and electromagnetism had shown that the speculations on the atomic-molecular-kinetic theory of matter were neither required nor, were in fact, possible with respect to a sound foundational basis of physical theory. Speculations about the atomicity of matter rapidly declined in importance until after the turn of the century. Even physical chemists sought more easily demonstrable assumptions about the physical world.[100] The idea that physics must make its goal the mechanical explanation of nature, therefore, did not go unchallenged. Among all the reactions to the mechanist-materialist dogma, a small group of ideas grew out of the dissatisfaction with classical physics (i.e., mechanics) and were to have an important influence on the philosophical interpretation and methodological approach of science

---

[99] See ibid.; S. G. Brush, "Thermodynamics and History: Science and Culture in the 19th Century," *Graduate Journal*, 7 (1967), 477–565, reprinted as *The Temperature of History: Phases of Science and Culture in the Nineteenth Century* (New York: B. Franklin, 1978); E. N. Hiebert, "The Energetics Controversy and the New Thermodynamics," in D. H. D. Roller, ed., *Perspectives in the History of Science and Technology* (Norman: University of Oklahoma Press, 1971), pp. 67–86.

[100] G. M. Fleck, "Atomism in Late 19th Century Physical Chemistry," *Journal for the History of Ideas*, 24 (1963), 109–10.

in the twentieth century. They included positivism, empirico-criticism, and energetics.[101] The one feature that all these positions had in common was a fundamental emphasis on a purely phenomenological viewpoint: Scientific theories should strive for economy of thought, rather than explain phenomena in terms of unobservables.[102] As Albert Einstein was later to put it: "One got used to operating with these [electromagnetic] fields as independent substances without finding it necessary to give oneself an account of their mechanical nature; thus mechanics as the basis of physics was being abandoned, almost unnoticeably because its adaptability to the facts presented itself finally as hopeless."[103] The transition from mechanistic to mathematical physics was the result of this general "positivistic" doctrine in science during the latter years of the nineteenth century.[104] Mathematics, rather than mechanics, became the ultimate model for the description of nature.

Nearly all of the pioneers in classical statistical astronomy had been trained first as mathematical physicists. The question of whether or not they were mechanistic materialists is not strictly relevant. What is clear, however, is the fact that they expressed the general positivistic outlook of their age in their scientific work. The emphasis was on mathematics and the framing of mathematical models. Although they were interested in the placement, position, distribution, brightness, and velocity of stars, they increasingly emphasized the analytic forms of these relationships and not the underlying reality of such macrophenomena. To a certain extent it is true that the nature of their subject matter precluded discussion of the mechanical explanation of nature. But it is also true that they fully incorporated the acceptable procedures of their age and sought explanations in terms of mathematical concepts and the observable entities of stellar astronomy.

Seeliger, Schwarzschild, Charlier, and Eddington were the

---

[101] For a discussion of these various scientific dogmas, see Brush, "Thermodynamics and History," pp. 527–31.

[102] Ibid., pp. 522–3.

[103] A. Einstein, "Autobiographical Notes," in P. A. Schilp, ed., *Albert Einstein: Philosopher-Scientist* (Evanston, Ill.: Library of Living Philosophers, 1949), pp. 25–7.

[104] See P. Frank, "The Mechanical Versus the Mathematical Conception of Nature," *Philosophy of Science*, 4 (1937), 41–74.

principal advocates of a mathematical basis to stellar statistics, but they differed from one another: although Charlier was a statistician and Schwarzschild and Eddington were theoreticians, Seeliger was neither. Schwarzschild's primary interest was in advancing the theoretical tools of a given discipline, not in writing exhaustive memoirs, wringing his subject dry.[105] Consequently, he would outline a fundamental theorem or develop a general method and leave its application to others. In contrast, although Seeliger might develop a new approach, such as that embodied in his fundamental equations, he would carry its implications to the very end. Thus in his last investigation of statistical cosmology, for example, Seeliger employed the fundamental equations, Schwarzschild's fourier method, and even Kapteyn's various numerical approaches in his fundamental quest for a *cosmology* of the Universe. Eddington, of course, plowed new territory in stellar motions using, as did Schwarzschild, whom he particularly appreciated, various statistical and theoretical approaches.

[105] For example, in 1912 Schwarzschild derived twelve integral equations involving the density, luminosity, and velocity relationships and combined them with various empirical functions (e.g., star-counts, mean-parallaxes, mean-proper motions). His analysis easily represents the most concise discussion of how far a mathematical theory could be extended into the development of stellar statistics as a mathematical science. See K. Schwarzschild, "Zur Stellarstatistik," *Astronomische Nachrichten*, 190 (1912), 361–76.

6

STATISTICAL COSMOLOGY AS A
RESEARCH PROGRAM, 1915–1922

During the classical period of statistical cosmology, roughly from
1890 to the early 1920s, the sidereal problem represented a fundamental research program, it was thought, that would provide
detailed knowledge of the form, structure, and architecture of the
stellar universe in terms of several very basic universal relationships. Just as Kepler had derived empirically three planetary laws,
so too did statistical cosmologists believe they would obtain stellar
laws (also three in number) as true of the galactic system as
Kepler's laws are about planetary motions. But these laws were
not thought to be mere statistical regularities, but actual laws of
nature. As Kapteyn and his colleague Pieter van Rhijn (1886–1960)
put it concerning their 1920 derivation of the luminosity relationship: "It is difficult to avoid the conclusion that we have here
to do with a law of nature, a law which plays a dominant part
in the most diverse natural phenomena."[1]

STATISTICAL COSMOLOGY, 1915–1920

The mathematical approach to statistical astronomy heightened
belief in the exactness of the lawlike nature within stellar studies.
The one stellar relationship that could not be empirically inferred
for the entire galactic system from stellar data alone was, of
course, the density law, which, in the words of Charlier, was
"so eagerly sought for these last years in the discussion of the
constitution of the Milky Way."[2] Interest in the mathematical

[1] J. C. Kapteyn and P. van Rhijn, "On the Distribution of the Stars in Space
Especially in the High Galactic Latitudes," *Mount Wilson Observatory,
Contributions,* no. 188 (1920), p. 13, reprinted in *Astrophysical Journal,* lii
(1920), 23–38.
[2] C. V. L. Charlier, "Studies in Stellar Statistics – The Distances and the Dis-
tribution of the Stars of the Spectral Type B," *Meddelanden Fran Lunds
Astronomiska Observatorium* (Serie II), 14 (1916), p. 28 (108 pp.) (hereafter
Lund Medd.).

derivation of a general density relationship did not abate until the mid-1920s when numerical approaches in statistical astronomy began to gain influence as the mathematical approach became increasingly sterile. The Dutch-American astronomer Bart Bok, who had been involved in some of the early work, recently reminisced about the climate of opinion dominating the early 1920s:

> Those of us who were at that time interested in the structure of our Milky Way System were from the start imbued with a deep respect for advanced mathematical-statistical analysis. I, for one, had a great admiration for the intricacies of Gaussian integrals, and I truly hoped that some day I might stumble on a solution that would give the world the one and only, definitive, mathematical form for *the* density function of our galaxy.[3]

Ignoring for the moment Kapteyn's various *numerical* solutions to this problem, before 1920 several major attempts were made *analytically* to characterize this relationship in general terms: Seeliger's integral equation solution in 1898, Schwarzschild's 1910 approach resulting in a quadratic density function (later used by Kapteyn in his 1920 solution), Eddington and Dyson's 1913 solution based on stellar motions, and a densely abstract solution in 1916 by Charlier.[4]

During the early years of the century, certain developments in statistics led to the belief that the entire theory of frequency curves and correlations, the types used extensively by Charlier, Kapteyn, Eddington, and others, could be reduced to a problem in integral equations.[5] Consequently, many hoped that the entire study of stellar statistics would become merely a branch in the theory of these equations. In applying the mathematical theory to the derivation of the density relationship, however, certain unacceptable conclusions emerged. For instance, the density

---

[3] Bart Bok, "Harlow Shapley and the Discovery of the Center of Our Galaxy," in Jerzy Newman, ed., *The Heritage of Copernicus* (Cambridge, Mass.: MIT Press, 1974), p. 51.

[4] References to Seeliger, Schwarzschild, Eddington, and Dyson have already been provided in Chapter 5. For Charlier, see "Studies in Stellar Distribution of the Stars of the Spectral Type B," *Lund Medd.* (II), 14 (1916).

[5] Arne Fisher, *The Mathematical Theory of Probabilities and its Application to Frequency Curves and Statistical Methods* (New York: Macmillan Co., 1922, 2nd ed.), p. 187.

relationship obtained independently by Seeliger and by Eddington and Dyson contained an *inverse* distance term, and because the physical center in virtually all models of the Universe developed before 1920 corresponded to the neighborhood of the Sun, an *inverse* distance implied an *infinite* density at the Sun itself. Conversely, Schwarzschild's solution for the density relationship led to a *zero* density at the galactic center, a result that was again inconsistent with empirical observations. Subsequent attempts made to address these problems resulted in various ad hoc additions to the density relationships.

To complicate matters further, as noted earlier in Chapter 5, problems dealing with the irregular density distributions emerged from considerations of the spectral nature of stars. Between 1912 and 1916, Seeliger, Kapteyn, Eddington, Charlier, and others investigated the distribution of the stars of various spectral types.[6] On writing to Walter S. Adams at the Mount Wilson Observatory in 1912, Kapteyn noted:

Now that we begin to know something of the luminosity curve of stars of one particular class of spectrum – I have already derived such a curve for the Helium stars [B spectral type] and will shortly try to do so for the A stars. . . . Before we can really . . . attack the question of the arrangement of stars separately for the different spectra, such a knowledge [of the colors] seems absolutely necessary. In my mind, the most important problem in sidereal astronomy would be: the study of the arrangement of stars in space (including star streams) *separately* for stars of different spectral type. . . .[7]

---

[6] See H. von Seeliger, "Ueber die Abhändigkeit der Verteilung der Sterns von verschiedenen Spektraltypen und der mittleren Parallaxen der Stern von der galaktischen Breite," *Astronomische Nachrichten*, 193 (1912), 161–76, H. von Seeliger, "Ueber die Verteilung der Sterne von verschiedenen Spektral-typen," *Astronomische Nachrichten*, 194 (1913), 137–43, A. S. Eddington, "The Distribution of the Spectral Classes of Stars," *Observatory*, 36 (1913), 467–71, J. C. Kapteyn, "On the Individual Parallaxes of the Brighter Galactic Helium Stars in the Southern Hemisphere, together with Considerations on the Parallax of Stars in General," *Astrophysical Journal*, 40 (1914), 43–126, Charlier, "Studies in Stellar Statistics – the Distances and Distribution of the Stars of the Spectral Type B."

[7] J. C. Kapteyn to W. S. Adams, 11 November 1912 (Hale).

Published in 1914, this work on the helium stars examined both the density and luminosity relationships. After discussing Schwarzschild's empirical formula and its analytic form, Kapteyn concluded that "the luminosity-curve of the ... [helium] stars is thus found to be a Gaussian error-curve...."[8] Even though it was assumed that the analytic form of the density and luminosity relationships across all spectral classes would be identical, the various parameters of the frequency functions would differ from one class to another. As he wrote to Hale in 1916 concerning the extension of this problem:

> [W]e can find the distributions in space of nearly all the Helium stars [and] ... there is a gradual transition in every direction from the Helium stars to the other types.... All this finished I will have to come to the A stars, which in the main I find to behave like the Helium stars. If I finish them too I think I may hope to solve the many riddles that remain for the rest.[9]

By the mid-1920s, problems with the density relationship in general led to more of a criticism of the elegant mathematics. The mathematical theory of stellar statistics required that the empirical data be smoothed out in order to preserve the mathematical necessity for *continuous* functions. Thus any irregularities in the distribution of stars in the stellar system were essentially ignored; otherwise the analytical expressions could be neither easily integrated nor described as relatively simple functions with a statistical "mean" and "dispersion." The fact of the matter was that the variation of the stars in terms of apparent magnitude and galactic latitude and longitude could not be expressed as a simple *general* analytical function.

Finally, there was a lack of consensus about the mathematical form that various frequency functions – particularly the luminosity, mean-parallax, and star-count relationships – should possess. With the exception of Kapteyn's certainty about his (1920) luminosity relationship, ambivalence had predominated in the

[8] Kapteyn, "On the Individual Parallaxes of the Brighter Galactic Helium Stars," pp. 82–3.
[9] J. C. Kapteyn to G. E. Hale, 26 March 1916 (Hale).

preceding years and by the mid-1920s had once again become common. In 1926 Charlier expressed the dominant attitude:

> Regarding the frequency function ... of the absolute magnitudes [luminosity law], the observations are not sufficient at present to make possible a direct determination of its analytical form. We have reason to believe it is nearly of the normal form, but it is very possible that a coefficient of skewness, as well as of excess, has a noteworthy value. Until further investigations clear up the question we may use the normal frequency form even of this function.[10]

The mean-parallax function, which formed the basis of so much of Kapteyn's work, was also of questionable status largely because Kapteyn's 1901 derivation remained its sole empirical source for nearly two decades. Also, because this function was reasonably secure only within the local solar neighborhood, Eddington, among others, believed that both numerical and analytical approaches would be "modified in an important degree, by a modern determination of the mean parallaxes of stars of different magnitudes."[11] It was not until 1923 that van Rhijn redetermined the mean-parallax function using much improved data.[12]

SEELIGER'S COSMOLOGY

Although there is a lengthy history of speculations about the detailed form of the Milky Way system, these theories all remained for Seeliger unclear and unconvincing.[13] At least as late

[10] C. V. L. Charlier, *The Motion and Distribution of the Stars* (Berkeley: University of California Press, 1926), p. 98.

[11] A. S. Eddington, *Stellar Movements and the Structure of the Universe* (London: Macmillan & Co., 1914), p. 230.

[12] P. van Rhijn, "On the Mean Parallax of Stars of Determined Proper Motion, Apparent Magnitude and Galactic Latitude for Each Spectral Class," *Groningen Publications*, 34 (1923), 105 pp.

[13] See Part I of this book for various speculations about the form of the Milky Way system, particularly those by Struve and the Herschels. For a contemporary view, see C. Easton, "On a New Theory of the Milky Way," *Astrophysical Journal*, 12 (1900), 136–58, and Seeliger's response "Remarks on Easton's Article 'On a New Theory of the Milky Way'," *Astrophysical Journal*, 12 (1900), 376–80.

as 1911, Seeliger was of the opinion that the data were wholly lacking in admitting conclusive results. As he described his understanding of the state of knowledge at the time to his lifelong friend Max Wolf (1863–1932), director of the Observatory at Heidelberg:

> You apparently believe that the width of the wide spaces [of nebulae] should correspond to the Milky Way system, but there is no proof for this. It might be acceptable if we knew that the spiral nebulae were of the same size as the Milky Way system. You have discovered yourself that the sizes of all spiral nebulae are not more than 1/20th of its size [the Milky Way], which in itself is a contradiction. Besides we could adopt the principal of homogeneity [i.e., uniformitarianism] and be more right. . . . Whatever one wants to assume may be correct; it would all be supposition.[14]

Investigations of the luminosity and stellar density laws, whether by Seeliger, Kapteyn, Schwarzschild, Charlier, Eddington, or Dyson, were normally treated together in complementary terms rather than separately, because both relationships were needed to describe the spatial distribution of the stars. Hence virtually all stellar astronomers considered these relationships as "the two fundamental laws which determine the arrangement of the stars in space."[15] A mathematical (or numerical) description of the spatial arrangement of the stars together with their total number was considered to constitute a complete solution to the sidereal problem of statistical cosmology.

The year 1920 was particularly rich for statistical cosmologists and for stellar astronomy generally. Within the first three months of that year both Seeliger and Kapteyn published their most mature, sophisticated models of the stellar universe. Moreover, both relied on a lifetime of accumulated knowledge and results, and they brought to bear techniques – analytical and numerical – that they and others had been developing for nearly half a

[14]   H. v. Seeliger to M. Wolf, 23 January 1911 (Heidelberg).
[15]   Kapteyn and van Rhijn, "On the Distribution of the Stars in Space Especially in the High Galactic Latitudes," p. 29.

century. As it turns out, their models incorporated the same principles and procedures that they introduced in their earliest investigations of 1898 and 1901, respectively. Of course, these newer developments reflected many advances achieved in both data collection and analysis and in stellar theory since the turn of the century.

Between 1898 and 1920 Seeliger produced five major treatises and nearly a score of additional scientific papers dealing directly with the sidereal problem.[16] In these published writings we can identify four basic concerns that were addressed in his investigations: the density law, the luminosity law, interstellar absorption, and the finiteness of the stellar universe.

With one exception, Seeliger basically used the same density function, initially derived in 1898, throughout his career. Using Kapteyn's later calculations of mean-parallaxes (1901) and star-counts (1908), Seeliger changed his density function slightly (in 1911) to accommodate the apparent constant stellar density within the immediate local solar region (within 10 parsecs). With this one minor change, though, Seeliger's density law closely followed the predictions based on his earliest views.

In 1898 Seeliger had become the first theoretician to succeed fully in incorporating a generalized brightness relationship into a theory of the spatial distribution of the stars. But throughout his career, Seeliger found great difficulty in analytically deriving a luminosity function completely consistent with the empirical data. Although his luminosity function was always Gaussian in nature, in his first model (1898) he assumed an entirely arbitrary probability function. In 1909 Seeliger used the luminosity relationship suggested by George C. Comstock, only to discover later that Comstock's function had been derived by a simple – though

---

[16] Seeliger's five major treatises in statistical cosmology, in which he laid out his theory of the stellar system, are "Betrauchungen über die räumliche Verteilung der Fixsterne," *München Ak. Abh.*, 19 (1898), 565–629; "Betrauchungen über die räumliche Verteilung der Fixsterne," *München Ak. Abh.*, 25 (1909), 2–53; "Ueber die räumliche Verteilung der Sterne im schematischen Sternsystem," *München Ak. Sber.*, 41 (1911), 413–61; "Ueber die räumliche und scheinbare Verteilung der Sterne," *München Ak. Sber.*, 42 (1912), 451–509; and "Untersuchungen über das Sternsystem," *München Ak. Sber.*, (1920), 87–144.

critical – misunderstanding of Kapteyn's results.[17] In his 1911 paper, where he first modified the density law, he corrected this earlier error and introduced an exponential form to the luminosity relationship.[18] Using a form that is mathematically equivalent to Charlier's statistical representation of the luminosity equation, Seeliger later changed it yet again for his 1920 model.[19] In his 1920 model he used these final functional representations in order to compare the theoretical calculations with the corresponding empirical data for the mean-parallax values. Doing so provided the crucial check on the validity of his equations.

As we have clearly seen, Seeliger's methods were dominated by a severe analytical approach. As he himself advocated, "the fundamental problems of stellar statistics are reduced to complicated problems in the theory of integral equations."[20] Ironically, it was the extreme complexity of the mathematics and the inability to resolve the observational data that eventually caused the almost total demise of the analytical approach. Nevertheless, Seeliger's 1920 theory was expressed in the rarefied atmosphere of mathematical theory. In order to reflect the extreme complexity within the empirical data, however, over time Seeliger was able to relax various simplifying assumptions on the sidereal problem. For example, whereas his earlier models (1909 and 1911) ignored the variation of stellar density with respect to galactic latitude, the 1920 theory included a discussion of this problem. Again in his later theories, Seeliger went on to treat the crucial problems of interstellar light absorption and the physical boundaries of the stellar universe using the mathematics of stellar statistics.

From his earliest studies he paid considerable attention to the

[17]   G. C. Comstock, "The Luminosity of the Fixed Stars," *Astronomical Journal*, 25 (1907), 169–73. See Seeliger, "Betrauchungen über die räumliche Verteilung der Fixsterne," (1909), pp. 19–22; Seeliger first adopted Comstock's luminosity function in his "Ueber die räumliche Verteilung der Sterne," *Astronomische Nachrichten*, 182 (1909), 236–8 (229–48). For Kapteyn's devastating critique of Comstock, see J. C. Kapteyn, "The Luminosity Curve," *Astronomische Nachrichten*, 183 (1910), 313–14 (313–32).

[18]   Seeliger, "Ueber die räumliche Verteilung der Sterne im schematischen Sternsystem," pp. 437–8. Also, see Seeliger, "Ueber die räumliche und scheinbare Verteilung der Sterne," pp. 457–9.

[19]   Seeliger, "Untersuchungen über das Sternsystem," (1920), p. 128.

[20]   Ibid., p. 88.

possible existence of interstellar absorption because of its potential deleterious effects on his conclusions. Numerous astronomers had discussed this topic, but in terms of the number of research papers (and pages) Seeliger and Kapteyn easily head the list. In his 1909 theory, first discussed in 1898, Seeliger introduced an arbitrary absorption term in order to compensate analytically for its possible existence. In 1911, 1912, and again in 1920 Seeliger began his mathematical manipulations by assuming this "contingent absorption" term.[21] And in these major papers, he would always conclude his discussion with a consideration of the problem of interstellar absorption. Frequently, he would derive theoretical values with and without absorption for purposes of comparison with various empirical data. At most he obtained a general light-extinction *effect* of about one part in ten thousand; that is, star-light should be diminished only by a fraction of its full brightness. "Extinction," concluded Seeliger, "has almost no effect on the [perceived brightnesses of] the stars."[22] Despite Seeliger's and Kapteyn's efforts, however, the problem of interstellar absorption remained a peripheral and somewhat speculative topic until 1930.

Of greater importance for his solution of the sidereal problem, however, was a careful consideration of the form, structure, boundary, and size of the sidereal universe. Though similar but considerably more expanded than his 1898 theory, Seeliger's most sophisticated statistical cosmology (1920) entailed a Sun-centered ellipsoidal stellar system approximately 10,000 parsecs in the Milky Way and 1,800 parsecs toward the galactic poles. Graphically represented in Figure 16, the lines represent surfaces of equal stellar density, whereas the heavy line represents the boundary of the sidereal system beyond which space is theoretically almost entirely devoid of stellar objects. Stellar density is largely equally uniform longitudinally, diminishing rapidly toward the poles.

Assiduously avoiding speculation, Seeliger's 1920 statistical model of the stellar universe followed developments he had already outlined in his 1898 theory. To be sure, he utilized the most sophisticated advances in analytic stellar statistics, always basing his

---

[21] Ibid., p. 94.   [22] Ibid., pp. 141–4.

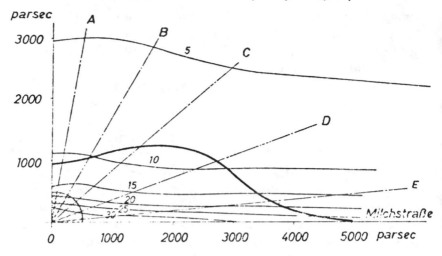

Figure 16 Seeliger's 1920 stellar system. The lines roughly parallel to the Milky Way – 30, 25, 20, 15, 10, and 5 – represent both planes of equal density and the degree (in percentages) of density vis-à-vis the Solar neighborhood (e.g., 30 = 30 percent of the density near the Sun). The lines A, B, C, D, and E refer, respectively, to the galactic latitudes 80°, 60°, 40°, 20°, and 5°.

investigations on the best empirical data available.[23] But his basic methodological approach, in terms of the fundamental equations, he had developed twenty years earlier.

THE "KAPTEYN UNIVERSE"

Although there are significant differences in Seeliger's and Kapteyn's methods, it has been suggested that "in principle there are no essential differences between their *conclusions* [italics added]."[24] This point was noted by others as well, and their work

[23] Seeliger published only one other paper after his 1920 stellar theory, and this was in response to a criticism of his data by van Rhijn. See P. van Rhijn, "Bemerkung zu Dr. H. Seeliger letzten 'Untersuchungen über das Sternsystem', "*Astronomische Nachrichten*, 213 (5091) (1920), 45–8; and Seeliger's reply "Bemerkungen zu dem Aufsatz des Herrn van Rhijn in A. N., no. 5091," *Astronomische Nachrichten*, 214 (1921), 145–50.

[24] H. Kienle, "Die räumliche Dichteverteilung im Sternsystem," *Naturwissenschaften*, 10 (1922), 684 (679–86).

suggests that statistical cosmology had achieved deep consensus during the early decades of the twentieth century.[25] More so than Seeliger's results, however, Kapteyn's 1920–22 theory of the stellar system became the epitome of investigations providing the classical solution to a *statistical* cosmology, and in the process came to be known as the "Kapteyn Universe," a term coined by James H. Jeans (1877–1946), the English mathematical astrophysicist.[26]

Occasionally, Kapteyn would share his ideas with a broader audience and reveal crucial issues that needed resolution. In an address before the National Academy of Sciences in April 1913, entitled "The Structure of the Universe," Kapteyn spoke on his star-streaming discovery in order to convey both the excitement as well as the frustration of his work. Far from offering some grand results, Kapteyn concluded "since the first man looked up to the sky, . . . a few pathways [have been] mapped out, along which we may direct a hopeful attack."[27] Elsewhere on star-streaming, he outlined the major technical issue needing resolution: "The problem of the present structure of the Universe is the problem of the distances. From the moment that the distances become known we shall be able to make a model which will be a true representation of our stellar system. . . . [T]he distances determine the present arrangement of the stars in space. . . ."[28] In essence, this is the problem first recognized by William Herschel a century-and-a-half earlier, and that, in the investigations of the

---

[25] R. J. Trumpler and H. F. Weaver, *Statistical Astronomy* (Berkeley: University of California Press, 1953), p. 438.

[26] The results of Kapteyn's latter work had been communicated at the Edinburgh meetings of the BAAS in September 1921. Because Jeans was present, he knew of the results prior to publication. Working independently, though relying on much of the same data, he reached basically the same conclusions as Kapteyn. See Jeans to Kapteyn, 28 December 1921 (University of Groningen Archives), and J. H. Jeans, "The Motions of the Stars in a Kapteyn-Universe," *Royal Astronomical Society, Monthly Notices*, 82 (1922), 122–32. For a discussion of Jeans's support of the Kapteyn Universe with his force studies and implications for the "island universe" theory, see Robert W. Smith, *The Expanding Universe: Astronomy's "Great Debate," 1900–1931* (Cambridge: Cambridge University Press, 1982), pp. 104–5.

[27] J. C. Kapteyn, "The Structure of the Universe," *Science*, 38 (1913), 724 (717–24).

[28] J. C. Kapteyn, "On the Structure of the Universe," *Journal of the Royal Astronomical Society of Canada*, 8 (1914), 145 (145–59).

statistical cosmologists of the early twentieth century, had become codified as the density function.[29]

Kapteyn's entire career was dominated – obsessed would be a better word – by his quest for a solution to the sidereal problem. His hoped-for solution would be fundamentally predicated, of course, on good, reliable data. Thus he was directly involved in mammoth projects for the collection of raw data, such as Gill's *Cape* survey, his own *Plan,* and later Pickering's photographic survey of the northern hemisphere.[30] His private correspondence, more so than his published work, indicates that he almost despaired of securing the data necessary to realize his goals. We could cite many examples from his letters, but one, written to Hale, will suffice: "In the study of the general system, however, we are still so little advanced and the data are so scanty that they are mostly not even sufficient for drawing up a good program."[31] This letter was written in 1918! Kapteyn was sixty-seven years old, and had been searching for, collecting, and analyzing data for nearly forty years, ever since he first gave his inaugural address at the University of Groningen in 1878.

His desire for a definitive solution to the sidereal problem was so compelling that, in order to obtain the needed empirical material, he continued to be directly involved in data reduction and analysis. Indeed, the vast bulk of his professional time was actually spent doing this sort of measuring work. Gill, Hale, Adams, and many others continually cautioned Kapteyn, however, that his talents would be better used if he would focus on synthesis and remove himself from "mere" cataloguing. As Hale expressed it in 1912:

> [A] man of your ability ought not to be compelled to devote time and attention to such a piece of routine [as measuring and data reduction]. The more opportunity

[29] Kapteyn and van Rhijn most fully addressed this problem in their "On the Upper Limit of Distance to which the Arrangement of Stars in Space can at Present be Determined with some Confidence," *Astrophysical Journal,* 55 (1922), 242–71.

[30] J. C. Kapteyn and E. C. Pickering, "Durchmusterung of Selected Areas between $\delta = 0°$ and $\delta = 90°$. Systematic Plan," *Harvard College Observatory, Annals,* 101 (1918), 456 pp.

[31] J. C. Kapteyn to G. E. Hale, 17 March 1918 (Hale).

you have for thought on the larger phases of astronomical work, the more will astronomy benefit through
the extraordinary range of your imaginative power.
Routine work may not do you any harm, but it will
certainly prevent you from dwelling on the larger theoretical aspects of the subject, and insofar as it does this
it will handicap you.[32]

By about 1915, in fact, Kapteyn was increasingly turning his
attention to these larger concerns. But he preferred to rely on the
older, tried-and-true methods that had served him well for so
many years. The time had finally arrived, he was convinced, for
the "grand synthesis." Thus with the assistance of his colleague
van Rhijn, in 1917 Kapteyn began his final onslaught on the
sidereal problem. Before his death in July 1922, he would obtain
some plausible results.

Throughout his entire career, Kapteyn always recognized the
absolute necessity of reliable data of the right sort, even if only
a provisional solution was to be achieved. Of course, the quality
and accuracy of the empirical data had increased immeasurably
since publication of his first stellar theory. In 1901 reliable data
were available only for stars brighter than sixth magnitude,
whereas by the late 'teens the data had been extended to the
twelfth magnitude, and began to be unreliable altogether only
beyond the fourteenth. In their now classic monograph on the
stellar universe (1920) in which they detailed their new theory,
Kapteyn and van Rhijn noted:

Now that, after so many years of preparation, out data
seem at last to be sufficient for the purpose ... of making possible an elaborate treatment of the arrangement of the stars in space, ... we have been unable

[32] G. E. Hale to J. C. Kapteyn, 3 December 1912. Also see Kapteyn's response
to Hale, 31 December 1912 (Hale), in which Kapteyn generally concurs
with Hale. For Gill's concern, see D. Gill to J. C. Kapteyn, 27 March 1907,
and D. Gill to E. C. Pickering , 25 October 1912 (K.A.L.), in which Gill
chides Pickering for soliciting Kapteyn's assistance. In correspondence with
Hale as early as 1905, Kapteyn expressed his devotion to the theoretical side
of these questions, but lamented the paucity of relevant data; see J. C.
Kapteyn to G. E. Hale, 7 May 1905 (Hale).

to restrain our curiosity and have resolved to carry through completely a small part of the work.[33]

Using Kapteyn's tabular method (1901), they derived from the newer data both the density law and the luminosity curve. The latter ranged over no fewer than eighteen magnitudes, and gave an approximation to the luminosities that was "astonishingly close." From their tabular results, the authors derived an analytic form for the new luminosity curve, the form of which Kapteyn had actually already discussed with Schwarzschild a decade earlier.[34] Not only did the curve as a whole approximate the empirical data, but most significant, in their view, both the maximum of the curve and its dispersion matched recently obtained data extremely well. Thus it was not necessary, as it had been in 1901, to extrapolate the curve in order to find the position of the Gaussian maximum. They labeled these results the "first solution" of their synthesis.

Although the luminosity curve was derived only for the region within a local solar neighborhood of a radius of 630 parsecs, within which it was assumed to be completely valid, it was obviously necessary to assume its *universal* applicability in order to deal with the *entire* stellar universe. Beyond 630 parsecs the paucity of directly observed data, such as parallaxes and proper motions, compelled a different approach. Here Kapteyn and van Rhijn employed Schwarzschild's version of the fundamental equation of stellar statistics, in which Schwarzschild accounted for both the luminosity curve and the star-count function using second-order approximations. Kapteyn and van Rhijn's work also yielded nearly identical second-order results. Because their solutions were in accord with the mathematical work of Seeliger and Schwarzschild, Kapteyn and van Rhijn felt confident that the new density law correctly described the actual state of affairs beyond the local solar region. The Seeliger–Schwarzschild mathematics was particularly useful at those distances for which neither the luminosity nor the density functions could be derived using Kapteyn's

[33] Kapteyn and van Rhijn, "On the Distribution of the Stars in Space Especially in the High Galactic Latitudes," pp. 23–4.
[34] J. C. Kapteyn to K. Schwarzschild, 11 April 1910 (Schwarzschild).

tabular, or numerical, approach of 1901. Because the star-counts were empirically known and the luminosity curve now derived anew, the fundamental equation analytically yielded the density function.[35] Triumphantly they concluded that with "astonishing approximation" their results confirmed Schwarzschild's mathematical results: "[T]his must give a first insight into the arrangement of the stars of the *whole* stellar system in space."[36] For purposes of their discussion, they called the derivation beyond 630 parsecs the "second solution."

By combining both the first and the second solutions, Kapteyn and van Rhijn completed their most comprehensive (static) model of the stellar universe (see Figure 17).[37] For distances within 100 parsecs the second solution gave unacceptable values, whereas the first solution gave values in accord with empirical observations. Between 100 and 630 parsecs the two solutions agreed surprisingly well and, consequently, the mean of the two was chosen. Beyond 630 parsecs only the second solution was valid. In the 1920 paper, they described a transparent, ellipsoidal stellar system in which star density at low galactic latitudes diminishes in all directions with increasing distance from the Sun-centered system. Overall the dimensions of the system were 2,400 parsecs toward the galactic poles and 18,000 parsecs in the galactic plane. At 600 parsecs star-density was about 60 percent of that near the Sun; at 1,600 parsecs about 20 percent; at 4,000 parsecs only 5 percent; and at its perimeter, about 9,000 parsecs from the Sun, star-density was less than 1 percent of the solar region. At high galactic latitudes, Kapteyn's results closely represented the observational data. Although Kapteyn and van Rhijn still admitted their solution to be no more than "provisional," many others

[35] See K. Schwarzschild, "Ueber die Integralgleichungen der Stellarstatistik," *Astronomische Nachrichten*, 185 (1910), 85–6 (81–8).

[36] Kapteyn and van Rhijn, "On the Distribution of the Stars in Space Especially in the High Galactic Latitudes," pp. 34–5.

[37] Ibid. This paper describing their model was simultaneously published in both of Hale's publications, the *Mount Wilson Observatory, Contributions* and the *Astrophysical Journal*. Many of the technical details were given more fully in J. C. Kapteyn and P. van Rhijn, "The Number of Stars between Definite Limits of Proper Motion, Visual Magnitude and Galactic Latitude for each Spectral Class, together with some other Investigations," *Groningen Publications*, no. 30 (1920), 110 pp.

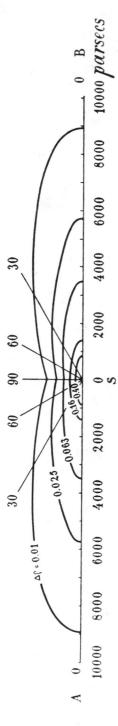

Figure 17 Kapteyn's 1920 stellar system. The curves represent lines of equal density distribution perpendicular to the plane AB of the Galaxy. The numbers 0, 30, 60, and 90 represent galactic latitudes in degrees. Density numbers are relative, with the density of the Sun (assumed to be at the center) taken as unity. (Reprinted with permission from *The Astrophysical Journal*)

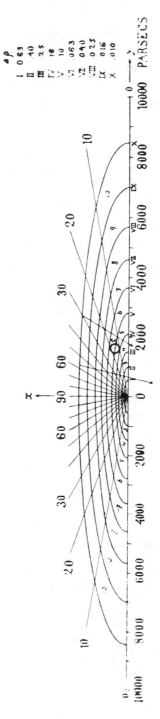

Figure 18 The Kapteyn Universe. The ellipsoids of revolution represent lines of equal density surfaces distributed perpendicularly to the plane of the Galaxy. Marginal numbers represent galactic latitudes; roman numerals represent (relative) densities with the center of the system taken as unity. Note the eccentric location of the Sun. (Reprinted with permission from *The Astrophysical Journal*)

considered their results nearly complete. Additional empirical data would refine only details, and not their general results. In their classic book *Statistical Astronomy* (1953), R. J. Trumpler and H. F. Weaver were to hail Kapteyn's 1920 theory of the stellar system as "the final achievement of a lifetime of masterly statistical investigations."[38]

Kapteyn's 1920 model represents the culmination of his life's work on the sidereal problem. All the solutions that he had provided since his earliest 1901 paper had followed the same general approach. Like the others that preceded it, this model is entirely static, and thus it fails to account for large-scale motion of the stars as represented in his discovery of star-streaming. Because his understanding of star-streaming vis-à-vis his static sidereal model had barely progressed in the years since 1902 and because so many others had attached such importance to this phenomenon, Kapteyn, somewhat anticlimactically, published a dynamic theory of the stellar system in 1922. This theory attempted to explain stellar motions in terms of gravitational forces, in the context of his earlier static, density distribution system (see Figure 18). In order to account for the shape of his (1920) galactic system through dynamic considerations, Kapteyn assumed that within the plane of the stellar system there is a general rotation about the polar axis, with the two star-streams accounting for the motion. Centrifugal forces plus random motions were balanced by the gravitational field. The flattening of the system was accounted for by a sufficient velocity of rotation. Because the two star-streams interpenetrate at a relative velocity of 40 km/sec, Kapteyn suggested that the Sun must be about 650 parsecs from the center of the system in order to account for a linear velocity of 20 km/sec. Regarding his earlier Sun-centered system, Kapteyn now concluded that "it seems infinitely improbable that the sun is at the center."[39] Kapteyn first presented these results at an

[38] Trumpler and Weaver, *Statistical Astronomy*, p. 438. Also see G. E. Hale, "Professor Kapteyn's Investigations," *Mount Wilson Observatory, Reports*, no. 19 (1920), 254–5.

[39] J. C. Kapteyn, "First Attempt at a Theory of the Arrangement and Motion of the Sidereal System," *Astrophysical Journal*, 55 (1922), 302–28. Kapteyn had been working on the gravitational distribution of forces since at least early 1921; see J. C. Kapteyn to G. E. Hale, 4 April 1921, 30 July 1921, and 3 January 1922 (Hale).

informal gathering of specialists, which included Albert Einstein and James Jeans, in November 1921 at Leiden.[40]

STATISTICAL COSMOLOGY, 1920–1922

Clearly, some details of the systems developed by Seeliger and Kapteyn were different. This was a relatively minor point however, until Harlow Shapley later challenged their conclusions.[41] In an effort of reconciliation, Hans Kienle, one of Seeliger's most gifted students, expressed personally to Shapley in 1922:

> Seeliger's results, a limited system from 6,000 to 26,000 lightyears in diameter, should be very sympathetic to you. One needs only to suppose that Seeliger's star counts refers only to the "local cluster,". . . while Kapteyn's "unlimited stellar system" with a density up to 40,000 lightyears distance does not appear to me to make much sense with respect to your stellar system. Kapteyn evidently realized this himself and accepted [Heber D.] Curtis' point of view. Especially if your opinion would prove to be correct, one would have to consider Seeliger's prominently outstanding boundaries of the local cluster, and [Kapteyn's] advantage over Seeliger would exist only in that this boundary is not to be measured with respect to the Milky Way. Rather it would point to a higher order in the system of the cluster.[42]

Although Kienle's motives were honorable, his criticism of Kapteyn, particularly in light of developments in statistical cosmology since the turn of the century, was exaggerated. Seeliger's and Kapteyn's theories were in fact comparable. After all, though they clearly favored one technique rather than another, they both used similar – if not identical – statistical and mathematical methods; and though they emphasized different data, they both relied largely on the same empirical information.

[40] H. Hertzsprung-Kapteyn, *J. C. Kapteyn: Zijn Leven en Werken* (Groningen: P. Noordhoff, 1928), pp. 172–3.
[41] See the discussion of Shapley's cosmology and the reaction to it by Kaptyen, Seeliger, and others in Chapter 8.
[42] H. Kienle to H. Shapley, 21 September 1922 (Shapley).

But these investigators were cosmologists! Their interest in large-scale astronomy was to answer fundamental questions. Kapteyn's 1920 model represented his lifetime achievements in dealing with the sidereal problem, whereas his 1922 theory was a provisional attempt to relate his 1920 model of the distribution of the stars with his earlier discovery of star-streaming. Seeliger's 1920 theory, though perhaps more restrictive and ultimately intended to be used for a critique of the inadequacies of universal gravitation, was to Kienle's mind a description of the local solar neighborhood that was ultimately compatible with Shapley's much larger universe. Both Schwarzschild and Eddington, though they never provided grand schemes of the stellar system, also contributed heavily to the methodological tools of statistical astronomy, eventually plowing new territory in relativity.[43] Charlier, who was younger than both Seeliger and Kapteyn, became increasingly sympathetic to Shapley's work. Therefore it was left to the founders of statistical cosmology, the Dutch, the Germans, and later the English and the Swedes, to refine the empirical support needed to explore the questions and investigate the mathematico-statistical basis of the sidereal problem.

[43] Even though Schwarzschild died in 1916, the year in which Einstein announced his "general theory of relativity," he and Eddington, both of whom had already become converts to Einstein's "special theory" (1905), began to understand their statistical studies in light of these newer, radical views. See A. S. Eddington to K. Schwarzschild, 14 September 1911 (Schwarzschild).

7

INTERNATIONALIZATION OF ASTRONOMY

Around the turn of the twentieth century, that is, roughly during the time when Kapteyn and Seeliger were most actively involved in cosmological developments, astronomy matured into a truly international enterprise. Even though by 1880 there still had been virtually no large-scale cooperative efforts in astronomical research, by 1920, despite the devastating effects of the First World War on international science in general, astronomy had nearly become the model of successful, productive, international cooperation in scientific research.[1] This is not to say that astronomers had not hitherto used the results and star catalogues that others had produced. Indeed, through the nineteenth-century stellar astronomers increasingly relied upon one another's data. And while others, principally the organizers of the Astronomische Gesellschaft and the Carte du Ciel, did in fact develop international connections, the successful cooperation of astronomers and observatories on a specific research project was first the creation of Kapteyn.[2] Although, for example, cooperative programs were established in solar research (by Hale) and spectroscopic binary stars (by Campbell) during the first decade of the twentieth century, by far the most successful program was undertaken by Kapteyn in stellar distributions.

The acquisition and analysis of data increasingly became a cooperative enterprise within the international community of astronomers, but Kapteyn was truly the first to introduce a single laboratory specifically designed for the analysis and reduction of

---

[1] See F. J. M. Stratton, "International Cooperation in Astronomy: A Chapter of Astronomical History," *Royal Astronomical Society, Monthly Notices*, 94 (1934), 361–72.

[2] See John Lankford, "The Impact of Photography on Astronomy," in Owen Gingerich, ed., *Astrophysics and Twentieth-Century Astronomy to 1950, Part A* (Cambridge: Cambridge University Press, 1984), vol. 4, 29 (16–39).

observational data derived from the results of other observatories. As the French astronomer Jules Baillaud noted in his opening remarks before the 1922 International Astronomical Union in Rome: "The three things that have revolutionized astronomy in the last half century are photography, telescopes, and [Kapteyn's] Laboratory in Groningen."[3]

Although major centers for astronomical research had emerged throughout the world roughly by 1900,[4] they were particularly well focused in Holland with Kapteyn, in Germany with Seeliger and Schwarzschild, in Sweden with Charlier, in England with Eddington and Dyson, and in the United States with Pickering, Campbell, and Hale. For my purposes, these astronomers are important because (1) they contributed to what became the dominant pre-relativistic cosmology (espoused principally by the Europeans), (2) many of them mentored scores of important second-generation astronomers, and (3) they established successful astronomical facilities. In the years immediately following 1900 these astronomers developed well-defined centers of research: Critical problems became clearly identified, methodological approaches were developed, dedicated research teams were assembled, and formal means for distributing results emerged. Furthermore, each of these defining characteristics was reflected in a unique style within the respective research communities.[5] Here we will deal with only those developments within those communities that concerned themselves primarily with the architecture of the sidereal universe using the approach defined by *statistical cosmology*. Some groups, such as Pickering's and Campbell's which undertook significant and relevant work, did

---

3 Quoted in H. Hertzsprung-Kapteyn, *J. C. Kapteyn: Zijn Leven en Werken* (Groningen: P. Noordhoff, 1928), p. 74. In addition to the three events Baillaud noted, the development of spectroscopy should be added to his list.

4 For recent studies of astronomical facilities, see Owen Gingerich, et al., "Astronomical Institutions," in Gingerich, ed., *Astrophysics and Twentieth-Century Astronomy to 1950: Part A*, pp. 111–33, Owen Gingerich, et al., "Two Astronomical Anniversaries," *Journal for the History of Astronomy*, 21(1) (1990), 1–153, and Steven J. Dick, et al., "National Observatories: An Overview," *Journal for the History of Astronomy*, 22(1) (1991), 1–99.

5 Although this point is beyond the scope of this study, for a useful discussion of "national style," see Nathan Reingold, "The Peculiarities of the Americans or Are There National Styles in the Sciences," *Science in Context*, 4(2) (1991), 347–66.

not themselves contribute directly to a full and complete solution to this problem.

KAPTEYN AND INTERNATIONAL SCIENCE

Kapteyn's interests in the Milky Way shaped the careers of several generations of his students. Although he employed a number of assistants, Kapteyn produced only eight students. Among these, however, have been some of the profession's most creative astronomers, including Willem de Sitter (1901), H. A. Weersma (1908), Pieter van Rhijn (1915), and through van Rhijn, Jan Oort (1924), one of the twentieth century's greatest astronomers,[6] Bart Bok (1932), and others, all of whom made fundamental advances in modern astronomy.[7] Concerning Kapteyn's influence, Bok reminisced fairly recently in an oral interview for the American Institute of Physics: "In astronomy, I think we all [the Dutch] derived from Kapteyn. It all started with Kapteyn. Oort was Kapteyn's student. I was Oort's student. I was van Rhijn's student; van Rhijn was Kapteyn's student. Pannekoek was a great friend of Kapteyn's. So the rise of Dutch astronomy was entirely Kapteyn."[8] Except for de Sitter, who began his career as Kapteyn's assistant, but later developed a career in relativity and astrophysics, Weersma, van Rhijn, Oort, and Bok all pursued their early research programs within the broad outlines developed by Kapteyn.

But Kapteyn's influence extended far beyond his native Holland. Although his astronomical career was extremely colorful and even controversial at times, his efforts helped internationalize the astronomical community. Not only was Kapteyn a gifted scientist, but he also managed, as nearly all leaders do, to stimulate institutional cooperation. We have already noted Kapteyn's involvement with Gill's *Cape Photographic Durchmusterung*, and his own *Plan of Selected Areas* was perhaps the first truly multinational astronomical effort in which numerous observatories and astronomers cooperated. His long association with many

[6] See H. van Woerden, et al., eds., *Oort and the Universe: A Sketch of Oort's Research and Person* (Dordrecht: D. Reidel Publ., 1980).
[7] The parenthetical dates refer to the year in which their doctorates were conferred.
[8] "Bart Bok interview," American Institute of Physics (15 May 1978), 24.

astronomers, and with the Mount Wilson Observatory in particular, only further highlights Kapteyn's abilities to achieve great success with limited resources. Perhaps the clearest statement of Kapteyn's status in the astronomical community came shortly after he received the Bruce Medal from the Astronomical Society of the Pacific in 1913. Referring to Kapteyn's stellar studies, George Ellery Hale noted: "You must not suppose for a moment that there was any mistake made in awarding you the Bruce Medal. In my opinion, no astronomical work of the past generation has been more significant or important than your own, and it is a compliment to the other men who have received the Medal to claim them with you."[9]

Kapteyn was motivated in his lifelong astronomical research by two questions: What is the arrangement and distribution of the stars in space? and what are the systematic motions and space velocities of the stars? Answers to these issues, argued Kapteyn and most statistical cosmologists, would provide knowledge of the form and structure of the stellar universe. As we have seen, Kapteyn's devotion to these singular concerns explains the motivation behind his entire research efforts. Perhaps more so than any astronomer before him, Kapteyn vigorously encouraged international cooperation on whatever scale was needed to get the major questions answered.

We may speculate that Kapteyn's unusual international interests may be explained as the combination of an abundance of native talent with the *lack* of a good observational climate in Holland. Or perhaps Kapteyn simply wanted to organize all the world's resources for the solution of his *own* problems. In any event, Kapteyn's extraordinarily active international interests directly benefited the astronomical community at large, and likely encouraged others to participate in cooperative endeavors.

His joining with Sir David Gill in the production of the *Cape Photographic Durchmusterung* from 1886 to 1900 became the catalyst that convinced Kapteyn to develop his astronomical laboratory. In the process Kapteyn's laboratory became a sort of international clearing house for data reduction and analysis. Another major area of international cooperation was launched

[9]   G. E. Hale to J. C. Kapteyn, 18 February 1913 (Hale).

at the 1904 International Congress in St. Louis when Kapteyn proposed his *Plan of Selected Areas.* In his *Plan,* which was meant to optimize scarce resources worldwide so that no two research observatories would duplicate observational work, Kapteyn selected 206 stellar regions. Each area would be carefully catalogued for basic stellar data, such as magnitudes, motions, and spectra. From these data, Kapteyn hoped to construct the cosmology of the heavens. Thus a solution to the sidereal problem, it was hoped, would become feasible – within Kapteyn's lifetime, as he repeatedly urged. As Kapteyn expressed to Hale in 1905:

> My only wish is to get all the data mentioned in my plan [of Selected Areas] because all my work here the last ten years has utterly convinced me that, though we may perhaps come to the knowledge of the *probable* structure of the stellar system with the data of the brighter stars, we cannot demonstrate it without extensive data for the fainter ones.[10]

Hale was in the audience when Kapteyn discussed his work on the sidereal problem and presented his ambitious *Plan.*[11] It was here that Hale first recognized in Kapteyn another scientist with whom he could explore his own views on international cooperation.[12] Hale himself had his own agenda, however. He had invited astronomers from around the world to St. Louis, where in a meeting on astrophysics Hale presented his plans to standardize issues on the solar spectrum and routine observations of the

[10] J. C. Kapteyn to G. E. Hale, 7 May 1905 (Hale).

[11] For important background material on Hale, see Helen Wright, *Explorer of the Universe: A Biography of George Ellery Hale* (New York: Dutton & Co., 1966), and Helen Wright, et al., *The Legacy of George Ellery Hale* (Cambridge, Mass.: MIT Press, 1972), Donald E. Osterbrock, "Failure and Success: Two Early Experiments with Concave Gratings in Stellar Spectroscopy," *Journal for the History of Astronomy,* 17(2) (1986), 119–29, and Donald E. Osterbrock, *James E. Keeler: Pioneer American Astrophysicist and the Early Development of American Astrophysics* (Cambridge: Cambridge University Press, 1987).

[12] Hale's entire relationship with Kapteyn is described in a lengthy letter to Kapteyn's daughter; G. E. Hale to Henrietta Hertzsprung–Kapteyn, 13 August 1926 (Hale). For some details of their relationship, see Richard Berendzen, Richard Hart, and Daniel Seeley, *Man Discovers the Galaxies* (New York: Science History Publ., 1976), pp. 24–34, 70–96.

Figure 19 George Ellery Hale, ca. 1910. (Courtesy of Mount Wilson and Las Campanas Observatories, Carnegie Institution of Washington)

Sun. Those in attendance agreed to the formation of a society to promote solar research; at Oxford the following year, the International Union for Cooperation in Solar Research was established.[13] Shortly after St. Louis, Hale personally invited Kapteyn to organize a committee in Holland on solar research (Hale's specialty) so that all might benefit from international exchange. By July 1907, the Amsterdam Academy of Sciences, under Kapteyn's direction, established a cooperative program.[14]

About the same time, Kapteyn raised the question with Hale of organizing a congress for the study of sidereal cosmology using statistical techniques, and forming an international committee to oversee work on the Plan of Selected Areas.[15] With Hale's encouragement, the committee was formed by late 1907, and, with Kapteyn as chairman, included Hale, Gill, Pickering, Schwarzschild, Dyson, Adams, and the German astronomer K. F. Küstner. In response to Kapteyn's request for assistance, Hale was prepared to write the National Academy of Sciences in Washington to form a subcommittee on "selected areas" headed by either Kapteyn or Pickering. Although this never materialized, in 1913 the academy, celebrating its fiftieth anniversary, invited Kapteyn (and representatives from England, France, and Germany) to deliver addresses in order, as Hale expressed it, "to have your great work become better known among our [American] men of science. . . ."[16]

There were a number of directors at major observatories who committed resources to Kapteyn's *Plan,* including Gill at the Cape of Good Hope, Schwarzschild at Potsdam, Pickering at Harvard, and E. B. Frost at Yerkes. It was Hale and the resources at the

[13] David H. DeVorkin, "Community and Spectral Classification in Astrophysics: The Acceptance of E. C. Pickering's System in 1910," *Isis,* 72(261) (1981), 36–7 (29–49). In 1919, the International Union for Cooperation in Solar Research was united with the Astrographic Chart Conference to form the IAU; see W. S. Adams, "The History of the International Astronomical Union," *Astronomical Society of the Pacific, Publications,* 6 (1949), 5–12.

[14] G. E. Hale to J. C. Kapteyn, 27 May 1904, 8 June 1904, 21 July 1904, and J. C. Kapteyn to G. E. Hale, 31 July 1904 (Hale).

[15] See DeVorkin, "Community and Spectral Classification in Astrophysics: The Acceptance of E. C. Pickering's System in 1910," p. 38.

[16] G. E. Hale to J. C. Kapteyn, 10 January 1913 (Hale). The address that Kapteyn delivered at the National Academy of Sciences was "The Structure of the Universe," later published in *Science,* 38 (1913), 717–24.

Mount Wilson Solar Observatory, however, that particularly interested Kapteyn. Mount Wilson was then completing construction of the world's largest telescope: the sixty-inch reflector (completed December 1908). In January 1907, Hale invited Kapteyn to Mount Wilson. Within another year, under Hale's influence, the Carnegie Institution in Washington (the controlling organization of the Pasadena facility) offered Kapteyn a research associate position with a $1,800 yearly stipend, which Kapteyn held from 1908 until his death in 1922. By 1912 Hale was planning construction of the Hooker 100-inch telescope.[17] Both instruments would have observation time allocated to work on the *Plan,* as well as to other activities, of course.

During Kapteyn's yearly sojourns at Pasadena throughout most of his vacation months (roughly August to November) from 1908 to 1914, Kapteyn and his wife came to regard the tranquility in California as their "paradise" where their "beloved Mount Wilson" provided both satisfying friends and superb facilities.[18] While in transit between New York and the West Coast, Kapteyn would nearly always visit other American astronomers. Private correspondence shows that Kapteyn conferred with some frequency with Edward Pickering at Cambridge, E. B. Frost at Williams Bay (Yerkes Observatory), occasionally C. D. Perrine at Mount Hamilton (Lick Observatory), Frank Schlesinger at Allegheny Observatory, and in Washington, D.C., with Simon Newcomb, Robert S. Woodward, president of the Carnegie Institution, and Charles G. Abbot, secretary of the National Academy of Sciences.

Shortly after Kapteyn started to visit Mount Wilson, he began to recommend several European astronomers to Hale for work on the mountain. Among his compatriots, Kapteyn recommended one of his most promising students (soon to become a colleague), Pieter van Rhijn, as a "good mathematician, more theoretical than practical," a trait Hale never fully appreciated.[19] Kapteyn also

---

[17] G. E. Hale to J. C. Kapteyn, 8 January 1907, 14 March 1907, 15 June 1907, 23 December 1907; and J. C. Kapteyn to G. E. Hale, 17 February 1907, 12 June 1907, 21 June 1907 (Hale). Actually the 100-inch was finally tested and proved only in November 1917.

[18] Mrs. J. C. (E.) Kapteyn to Mrs. K. Schwarzschild, 1 May 1911, and 3 October 1913 (Schwarzschild).

[19] J. C. Kapteyn to G. E. Hale, 6 December 1911, 26 December 1911, 21 January 1912, 17 November 1912 (Hale).

encouraged Hale to consider Adriaan van Maanen, who had studied at Groningen between 1908 and 1911, and whose subsequent relationship with Mount Wilson and Edwin Hubble is legendary.[20]

Not only did Kapteyn recommend a number of Dutch astronomers, but he also recommended several Germans, indicating a closer relationship with Seeliger or at least with certain developments in German astronomy. All of the Germans Kapteyn recommended were or had been assistants at the Astrophysical Observatory at Potsdam.[21] The director at Potsdam was Karl Schwarzschild, whose brilliant theoretical work included his ellipsoidal theory of the star-streams. Along with Eddington's studies, Schwarzschild's work virtually assured Kapteyn's discovery a permanent place within the annals of astronomical history. Jan Oort, though never personally acquainted with Schwarzschild, reflected the attitudes of the Dutch toward Schwarzschild: "I became acquainted with Karl Schwarzschild in my Groningen student days amongst others through his visionary work on the ellipsoidal distribution of stellar velocities, and, again, later in Leiden, through the deep veneration with which Hertzsprung so often spoke of him."[22] Schwarzschild also had been Seeliger's finest doctoral student, and it is very likely he would have been one of the twentieth century's greatest astronomers (along with Eddington) had he not died prematurely in 1916.[23]

Kapteyn held Schwarzschild in such high regard that he recommended several of Schwarzschild's assistants to Hale. As a result, Hale came to respect the opinions not only of Kapteyn but also

---

[20] See, for example, Norris Hetherington, "Adriaan van Maanen and Internal Motions in Spiral Nebulae: A Historical Review," *Royal Astronomical Society, Quarterly Journal*, 13 (1972), 25–39; idem, "Edwin Hubble on Adriaan van Maanen's Internal Motions in Spiral Nebulae," *Isis*, 65 (1974), 390–3; and idem, *Science and Objectivity* (Ames: Iowa State University Press, 1988), chap. 8. Also see Robert W. Smith, *The Expanding Universe: Astronomy's "Great Debate," 1900–1931* (Cambridge: Cambridge University Press, 1982), pp. 97–111, 133–6.

[21] During the period of this study, Potsdam was the national observatory of Germany (and until recently the German Democratic Republic).

[22] J. H. Oort, "On the Problem of the Origin of Spiral Structure," *Mitteilungen der Astronomischen Gesellschaft*, 32 (1973), 15.

[23] Jan Oort considered Schwarzschild the greatest German astronomer since Kepler; see ibid.

Figure 20 Kapteyn speaking to Dr. McBride, with Karl
Schwarzschild (middle right) and Vesto M. Slipher (far right) at
Mount Wilson Observatory, 1910. (Courtesy of Yerkes Observatory)

of Schwarzschild. Writing in 1909 to express his congratulations
to Schwarzschild for the latter's election to the directorship of the
Astrophysical Observatory at Potsdam, Hale noted: "I am one of
many who have watched your rapid progress with sincere sat-
isfaction, and with great respect for your important work...."[24]
In 1911, Hale engaged Schwarzschild's assistant Arnold Kohlschütter for spectrographic work, a decision that proved mutually
beneficial. As Frederick H. Seares, a senior astronomer at Mount
Wilson, expressed personally to Schwarzschild in October of 1911:
"Dr. Kohlschütter has been with us for some weeks, and we are
enjoying him very greatly. He has fitted into our small colony on
the mountain admirably."[25] Kohlschütter himself remained at the
"mountain" for many years until his retirement. Hale's assistant

[24] G. E. Hale to K. Schwarzschild, 20 October 1909 (Schwarzschild).
[25] F. H. Seares to K. Schwarzschild, 12 October 1911 (Schwarzschild).

Walter S. Adams eventually took a great dislike to Kohlschütter (because he was German), but otherwise (because of the nascent camaraderie between Hale and Kapteyn) Hale and Adams continued to ask Kapteyn for recommendations. Kapteyn suggested Ejnar Hertzsprung (who became Kapteyn's son-in-law) and the German Hans Ludendorff, both of whom were also assistants of Schwarzschild, and of course van Maanen.[26] Though "not at all a 'cataloguer' but a man of ideas," Kapteyn strongly preferred Hertzsprung. Originally trained as an engineer, Hertzsprung lacked some experience, particularly in spectroscopy, an area Hale was looking to fill. Although "Schwarzschild will not be pleased to lose the man," wrote Kapteyn to Hale in 1911, Hertzsprung

> lives for astronomy – and though he certainly is not so experienced a spectroscopist . . . he certainly is an able man, . . . whose burning enthusiasm will soon make up for any want of experience as a spectroscopist. . . . I may as well say that Schwarzschild long ago told me that [Hertzsprung] is really the only man at Potsdam with whom he wholly sympathizes both as a man and as a scientist.[27]

Following the successful experience with Kohlschütter, Hale enthusiastically responded to Kapteyn in April of 1912: "Please let me know if [Hertzsprung] finds it impossible to get the necessary subsidy from the government, as I think I could assist toward the payment of his expenses if this should prove to be necessary."[28] Schwarzschild's work was already known internationally, and therefore Hale invited Hertzsprung as well as van Maanen, who worked for Kapteyn at Groningen (1909–10), along with van Rhijn, who was Kapteyn's own assistant, for the summer of 1912. Neither Hertzsprung nor van Rhijn were entirely satisfactory from Hale's point of view, but van Maanen, in Hale's

[26] J. C. Kapteyn to W. S. Adams, 26 March 1911; W. S. Adams to J. C. Kapteyn, 20 April 1911, 15 February 1912; J. C. Kapteyn to G. E. Hale, 14 December 1911, 10 March 1912, 6 December 1911; G. E. Hale to J. C. Kapteyn, 20 April 1911, 11 November 1911, 29 March 1912 (Hale).

[27] J. C. Kapteyn to G. E. Hale, 10 March 1912, 6 December 1911 (Hale).

[28] Hale's remarks are quoted in a letter from J. C. Kapteyn to K. Schwarzschild, 15 April 1912 (Schwarzschild).

words, was "as satisfactory as ever – a great addition to our staff, I believe."[29]

Kapteyn's internationalism is perhaps most evident in his private attitude toward the First World War. Both his private correspondence and his public statements clearly indicate that he was largely impervious to national and political interests that might otherwise have diverted him from his astronomical goals. This is not to say that he failed to understand differences in national style and temperament, but he viewed the scientific community as, above all, multinational – even supranational. Thus his allegiance really seems to have been far more inclined to the scientific community than to any national or ideological group. Although the war permanently interrupted his yearly sojourns at Pasadena, Kapteyn believed that something good might still emerge from the war – a "federated Europe," as he called it.[30] No doubt a united Europe, in Kapteyn's mind, would enhance scientific collaboration.

Of course, Hale was disappointed that Kapteyn could no longer visit Mount Wilson after the outbreak of the war. Although America remained ostensibly neutral, Hale lamented the impact of "this disastrous war" on Germany, England, France, and particularly on science.[31] Eventually, however, Hale became emotionally and intellectually involved in the war, becoming almost bitterly anti-German. Kapteyn, de Sitter, and other Dutch scientists followed the position of their native Holland and remained "neutral."[32] During the war, Kapteyn continued an active correspondence with Hale in America as well as with Schwarzschild in Germany, even visiting Potsdam on at least one occasion.[33]

[29] G. E. Hale to J. C. Kapteyn, 3 December 1912, 26 December 1911; and J. C. Kapteyn to G. E. Hale, 21 January 1912, 17 November 1912 (Hale).

[30] J. C. Kapteyn to G. E. Hale, 7 March 1915; G. E. Hale to J. C. Kapteyn, 9 March 1916 (Hale).

[31] G. E. Hale to K. Schwarzschild, 8 April 1915 (Schwarzschild).

[32] See, for example, Hale's correspondence with Arthur R. Hicks, 14 March 1921, and Willem de Sitter, 30 March 1921. As late as April 1915, however, Hale still considered himself neutral with regard to the war, as did most Americans, including President Woodrow Wilson; see G. E. Hale to J. C. Kapteyn, 12 April 1915 (Hale).

[33] See J. C. Kapteyn to K. Schwarzschild, 21 June 1915 (Schwarzschild), written while Kapteyn was visiting the observatory in Potsdam.

Throughout the war, some American, many British, and virtually all French and Belgian scientists were adamantly opposed to resuming relations with German scientists once hostilities had ceased.[34] After the Armistice was concluded, and as scientists began to repair the enormous intellectual and emotional damage inflicted on the scientific community, there gradually emerged resentment by some American and British astronomers toward some of the Dutch scientists. Three events, in particular, generated some hostility between Hale and Kapteyn that for a time dampened the international cooperative work of these two leading astronomers: the formation of the International Research Council (1918) and the International Astronomical Union (1921), and an upcoming celebration of Kapteyn's seventieth birthday in 1921.

As the war progressed, Hale, who was not as extreme in his anti-German feelings as the French and Belgians, increasingly lobbied for the formation of what eventually became known as the International Research Council (IRC), which, among other things, would replace existing international societies and coordinate newly organized scientific associations from among allied – and eventually neutral – countries. Those in support of President Woodrow Wilson's policy of conciliation toward the Central Powers, however, were unwilling to support any new scientific organization that would squeeze out Germany and Austria. Among the neutrals, Kapteyn and others were equally unwilling to do the same. In support of his apolitical view of international science, Kapteyn circulated a public letter of protest that eventually included 278 signatories drawn from many of the neutral countries.[35] Among the Allied victors, the French and Belgians were adamantly opposed to Kapteyn's appeal not simply because they were reluctant to forgive German sins during the war, but because they wished to protect themselves – and science in general – from any vengeful ambitions of the vanquished. By the summer of 1920, every neutral country on the list of signatories, including Holland, had joined the IRC.

---

[34] Daniel J. Kevles, " 'Into Hostile Political Camps': The Reorganization of International Science in World War I," *Isis*, 62 (1971), 47–60.

[35] Draft of letter, William W. Campbell to *Science* magazine, December 1919, Hale MSS, Box 61, NRC Division of Foreign Relations file.

Before this split between Hale and Kapteyn developed, how-ever, they solicited each other's assistance. Writing to Kapteyn in June 1919 on the question of the formation of the International Astronomical Union, Hale remarked: "[S]teps will be taken to initiate the proposed new International Astronomical Union. . . . It goes without saying that we shall immediately need your co-operation and that of the other Dutch astronomers in the work of the IAU. . . ."[36] In numerous letters to astronomers in England and Holland, particularly to Kapteyn (before he realized Kapteyn's "neutralist" views), de Sitter, Frank Dyson, and H. H. Turner, Hale expressed deep opposition to allowing German astronomers and scientists in general from joining these international organiza-tions. Responding to Hale's position and attempting to defuse a serious misunderstanding, de Sitter, also a neutralist during the war, "defended" German astronomers: "You must however not forget that there was not a single *astronomer* amongst those 93 ['Manifesto of 93']."[37] Forged above the signatures of ninety-three prominent German scientists and scholars, the manifesto, issued on 4 October 1914 in most of the major German news-papers and in ten languages, deplored the accusation that Germany was responsible for the conditions that eventually led to the conflict and the charges that the German forces had committed atrocities in Belgium. As the war progressed, the manifesto became infamous, and many of those who signed it, including Max Planck, the dean of German science at the time, were equally resented.[38]

[36] G. E. Hale to J. C. Kapteyn, 10 June 1919 (Hale).

[37] W. de Sitter to G. E. Hale, 8 December 1920 (Hale). The "Manifesto of 93" refers to a petition signed and circulated by ninety-three German scientists who pledged their support to the kaiser and Germany's "moral right" in the Great War. Many leading German scientists, including Seeliger, found such a politically motivated declaration anathema.

[38] "Manifesto of 93," in G. F. Nicolai, *The Biology of War*, trans. by C. A. Grande and J. Grande (New York: The Century Co., 1918), pp. xi–xiv; also see Klaus Böhme, ed., *Aufrufe und Reden deutscher Professoren im Ersten Weltkrieg* (Stuttgart, 1975). See J. L. Heilbron, *The Dilemmas of an Upright Man: Max Planck as Spokesman for German Science* (Berkeley: University of California Press, 1986), pp. 69–81. Many of those who signed the mani-festo, such as Max Planck, the doyen of German science, did so on the strength of the reputations of other signers and did not themselves read it thoroughly. For additional remarks concerning this episode, see Elisabeth Crawford, *Nationalism and Internationalism in Science, 1880–1939* (Cambridge: Cambridge University Press, 1992).

Efforts by some, including de Sitter, to honor Kapteyn with the publication of his works on his seventieth birthday led still others who were more likely to be involved emotionally in the war to resist, particularly after it was suggested that Kapteyn was neutral, if not pro-German. After Hale had become aware of these circumstances, he explained "Kapteyn's apparently pro-German tendencies" by reminding de Sitter:

> I was about to reply to your letter of December 31 when a new situation developed, which led me to consult Seares and [Charles Edward] St. John (both close friends of Kapteyn's) and then to cable you as follows: 'Receiving serious criticism of Kapteyn. Advise organize no committees outside Holland. Solicit foreign contributions only from personal friends.' We were led to this action by the receipt of information from England that Kapteyn's acceptance of the decoration [Order] Pour le Mérite from the Kaiser (at the same time that it was given to the commander of the submarine that sank the Lusitania); his circular letter regarding the resumption of personal relations with German men of science; and his attacks on the International Research Council and the IAU, which he refuses to join, have led to very sharp criticisms from astronomers asked by Dyson to join his [Dyson's] committee [for the Kapteyn celebration]. . . . But it is now certain that not only in England, but in other countries (probably including the US), Kapteyn's apparently pro-German tendencies will lead to much more criticism of the same kind if we go on with the original plan. . . . While I do not believe that Kapteyn was really pro-German during the war, and while I have no personal feelings against him because of his opposition to the International Research Council and the associated International Unions, I now recognize that his attitude will naturally arouse much feeling against him among some of those asked to subscribe toward an international tribute.[39]

[39]   G. E. Hale to W. de Sitter, 10 February 1921 (Hale).

De Sitter immediately responded to the charges made in Hale's letter by countering all the allegations, even suggesting that Kapteyn was "anti-German," and reminding Hale that Kapteyn was awarded the *Order Pour le Mérite* before submarine warfare began in August 1914: "Kapteyn was in America at the time, and did not accept it till after his return to Holland in September or October."[40] Indeed, Kapteyn initially declined this award, which some considered to be the most prestigious scientific award at the time, because he felt that Dutch neutrality had been violated. The German Consul in Groningen assured Kapteyn, however, that such was not the case, whereupon Kapteyn agreed to its bestowal.[41] Despite his comments to de Sitter (quoted earlier), Hale was still bothered, and continued to refer to Kapteyn, in private correspondence, as an "extreme idealist."[42] As a result, the international celebration of Kapteyn's seventieth birthday never materialized.

As a result, Kapteyn was not invited to assist in the formation of the International Astronomical Union nor did he participate in its first official meeting in Rome, Italy in May 1922. Under normal circumstances, Kapteyn would have been central to such activities. Nevertheless, Kapteyn continued his active participation in other international gatherings, attending the Deutsche Astronomische Gesellschaft meetings in Potsdam in August 1921 and the BAAS meetings at Edinburgh in September.

SEELIGER AND GERMAN ASTRONOMY

Seeliger had close ties to the Astrophysical Observatory at Potsdam, which eventually became preeminent among Germany's astronomical facilities. During a very successful career at the Munich Observatory, Seeliger himself was offered the directorship at Potsdam, but declined and recommended his star pupil Karl Schwarzschild, who became director in 1909 and continued so until his tragic and untimely death in 1916. Ironically, although Schwarzschild became an articulate spokesman for Einstein's theory, Seeliger never fully appreciated relativity. As explained

---

[40] W. de Sitter to G. E. Hale, 9 March 1921 (Hale).
[41] Hertzsprung-Kapteyn, *J. C. Kapteyn*, pp. 145–6.
[42] G. E. Hale to F. W. Dyson, 30 March 1921 (Hale).

earlier, Seeliger had attempted to account for perceived inadequacies in Newtonian gravitation theory by fiddling with the mathematics; however, his efforts were to no avail.

Although Kapteyn was a sometimes brilliant diplomat and organizer among diverse interests whose efforts to internationalize astronomy were almost singular, (and rivaled only by Hale's reception of foreign astronomers), Seeliger never achieved the international attention eventually accorded Kapteyn. Seeliger's approach to the technical problems entailed in (statistical) cosmology was very severe and rigorous mathematically, whereas Kapteyn's style seemed more intuitive and straightforward. At one point late in their careers, this difference prompted Kapteyn indirectly to become almost excessively critical of Seeliger's work. In 1918, Willem J. A. Schouten had completed a dissertation under Kapteyn's direction in which he vehemently defended his mentor's approach to statistical cosmology over the analytical developments suggested by Seeliger and others. In his dissertation, *On the Determination of the Principal Laws of Statistical Astronomy,* Schouten devoted nearly one-fourth of his entire work to a severe criticism of Seeliger's approach. Although it is true that Kapteyn directed Schouten's dissertation, Kapteyn refused to have it published in his *Groningen Publications,* and indeed strongly urged Schouten not to publish it at all. The dissertation is so excessively polemical that it is entirely understandable why Kapteyn did not want it published: It would have been an embarrassment to Kapteyn personally as well as to his larger efforts in astronomy. Still, Schouten persisted and published it despite Kapteyn's protests, suggesting, at least according to one account, that Schouten was a very difficult person and somewhat of a renegade student.[43]

As a result, Seeliger lost whatever interest he had in the less rigorous approach Kapteyn advocated.[44] In a personal letter dated 1923 to his lifelong friend and colleague Max Wolf, director

[43] "Bart Bok Interview," p. 33.
[44] W. J. A. Schouten, *On the Determination of the Principal Laws of Statistical Astronomy* (Amsterdam: W. Krichner, 1918). For Seeliger's seasoned reply see "Untersuchungen über das Sternsystem," *Sitzungsberichte der Mathematisch-Physikalischen Klasse der K. Bayerischen Akademie der Wissenschaften zu München,* 1 (1920), 94 (87–144).

of the Observatory at Heidelberg University, Seeliger revealed his disappointment with the Dutch: "My works on this subject [statistical methods] have been largely ignored on account of the degrading influence of Kapteyn and his lieutenants; I do hope that this will be recognized officially."[45] Kapteyn did not confer with any regularity whatsoever with Seeliger, but Seeliger was still Kapteyn's only scientific peer, and he was a force both to be reckoned with and to be respected. Although Kapteyn and Seeliger almost single-handedly championed statistical cosmology and although Kapteyn was close personal friends with several of Seeliger's former students, principally with Schwarzschild and later in life with Gustav Eberhard, who became the senior observer at Potsdam following Schwarzschild's death, they rarely corresponded with one another directly.[46]

As we have noted, Seeliger's style of research was considerably different than Kapteyn's. Above all, Seeliger's work is best characterized by a very severe, relatively abstract approach, combining the subject matter of observational astronomy with obtuse developments in mathematics. Nevertheless, even though Seeliger engaged in limited international projects, he produced a host of talented astronomers, thirty-four altogether. The most distinguished were Schwarzschild (1895), who was quickly becoming one of the world's leading authorities, Hans Kienle (1918; 1895–1975), who became Germany's leading astronomer after Seeliger's death in 1924, and Paul ten Bruggencate (1924; 1901–61), one of Seeliger's last students, who distinguished himself as a solar specialist and worked for a short time with Hale at Mount Wilson.[47] The Germans, except principally for Schwarzschild, Kienle, and ten Bruggencate, mostly had influence only in their native Germany. This despite the fact that Seeliger's students held directorships at various German observatories, including those at Leipzig (J. Bauschinger), Gotha (E. Anding), Hamburg (R. Schorr), Berlin-Babelsberg (K. F. E. Bottlinger), Potsdam (K. Schwarzschild;

---

[45] H. v. Seeliger to M. Wolf, 17 February 1923 (Heidelberg).
[46] The only extant letter between Kapteyn and Seeliger has been reproduced in Hertzsprung–Kapteyn, *J. C. Kapteyn*, p. 152, in which Kapteyn tells Seeliger that he will accept appointment to the steering committee of the Astronomische Gesellschaft meetings in Potsdam to be held in 1921.
[47] The years in parentheses refer to the years in which they received their doctorates under Seeliger's tutelage.

H. Kienle), Frankfurt/m (R. Hess), and Göttingen (P. ten Bruggencate).[48]

Except for three Americans, all of Seeliger's students were Germans. The Americans included G. W. Myers (1896), who was on the faculty at the University of Chicago, and B. L. Newkirk (1902), chief of research (1928) at the General Electric laboratory in Schenectady, New York. The third American was Henry Smith Pritchett (1895; 1857–1939), who became director of the observatory at Washington University (1883), superintendent of the U.S. Coast and Geodetic Survey (1897), president of MIT (1900), and president of the Carnegie Foundation for the Advancement of Teaching (1906).

As a student, Pritchett was deeply impressed with Seeliger, and Seeliger, who was already familiar with Pritchett's astronomical work at the Washington Observatory, found the younger American fully prepared to become a candidate for the Ph.D. within a year after Pritchett's arrival in Germany in 1894.[49] After his stay in Germany, Pritchett was the only one among Seeliger's American students who was positioned well enough to influence other American astronomers to understand the important work of his mentor. For example, shortly after Pritchett's return from Germany, astronomers generally supported Pritchett to fill the vacancy of the superintendency of the U.S. Coast and Geodetic Survey, a position of considerable significance within the scientific community. Letters of support, from Hale, E. E. Barnard, T. J. J. See, and other American astronomers, indicated that Pritchett, as "one of the most eminent astronomers of the continent," was uncommonly qualified: "To put the Coast and Geodetic Survey upon a proper basis we need at once a scholar of the best scientific training, and a gentleman commanding the respect not only of

[48] Seeliger continued actively to campaign on behalf of his numerous students in order to get them placed in influential positions in the German scientific community. See H. v. Seeliger to David Hilbert, 10 December 1894 and 26 January 1895 (Göttingen), H. v. Seeliger to Felix Kline, 10 July 1897 and 23 July 1897 (Göttingen), H. v. Seeliger to Max Wolf, 12 October 1898, 19 October 1898 (Heidelberg), H. v. Seeliger to Wilhelm Wien, 5 December 1906 (Deutsches Museum), H. v. Seeliger to Science Minister (Munich) 30 October 1918 and 27 January 1922 (B. H. A.).

[49] Abraham Flexner, *Henry S. Pritchett: A Biography* (New York: Columbia University Press, 1943), pp. 39–49.

the Government, but also of his associates and of the scientific and general public."[50] Unfortunately, when Pritchett assumed the superintendency in 1897 and permanently left the academic world of teaching and research for scientific administration, he was in a position to influence his favored astronomy only indirectly.

Even so, Pritchett regularly corresponded with Percival Lowell, A. E. Douglass, and Vesto M. Slipher, all Lowell Observatory astronomers, and while serving on the Board of Regents of the Carnegie Institution, which supported the Mount Wilson Solar Observatory, Pritchett carried on an extensive personal correspondence with Hale.[51] Although Hale's impression of Seeliger was certainly influenced by Pritchett, Pritchett never seems to have convinced Hale to develop scientific connections with the Germans as Hale had been doing with the Dutch. Hale most certainly was also influenced by Walter S. Adams's impression of Seeliger, because Adams, who became Hale's assistant in 1901 at Yerkes and after 1904 was Hale's second-in-command at Mount Wilson, spent a full year (1900) at the University of Munich, where he had considerable opportunity to interact with both Seeliger and Schwarzschild during doctoral studies under Seeliger's direction. Adams's year with Seeliger, however, left him unimpressed with Seeliger's rigorous approach to astronomy. As a result, Adams made it clear to Hale that he was willing to forego his graduate studies in Munich in order to return to America to work with Hale – even if only for a year or so.[52]

Whether intentionally ignored or not, Seeliger's influence, particularly here in America, appears to have been minimal. Harlow Shapley, who, as director of the Harvard University Observatory, became a major force in American astronomy after 1920 with the dissemination of his revolutionary studies in galactic structure, concurred with this view. In a 1922 letter to Hans Kienle, Seeliger's

[50] H. M. Paul to L. J. Gage, 13 October 1897; T. J. J. See to L. J. Gage, 9 September 1897, and G. E. Hale, et al., to L. J. Gage, 3 August 1897 (H. S. Pritchett Papers, Library of Congress).

[51] See Lowell Microfilm and Hale Microfilm, *passim*.

[52] W. S. Adams to G. E. Hale, 31 December 1900 (Yerkes). I am indebted to Ronald Brashear of the Huntington Library for bringing this letter to my attention. Paul ten Bruggencate, one of Seeliger's last doctoral students, worked for Hale briefly around 1929.

most prominent student after Schwarzschild died and Germany's most important astronomer of the twentieth century, Shapley revealed:

> I agree with you that we appear to know too little concerning the contributions made by Seeliger to the problem of galactic structure. There are three insufficient reasons for this. Seeliger's publications are not as accessible as some. We in America, at least, have been influenced and gained greatly by Kapteyn and his school, and they have not been quite fair, it now seems to me, to Seeliger's work.[53]

Kapteyn, Seeliger, and Hale all recognized for their own purposes the value of international cooperation. Even though Seeliger attracted few international students, he remained president of the prestigious Deutsche Astronomische Gesellschaft for twenty-five years (1896–1920), the only astronomical society before the war that attempted to maintain a serious international profile.[54] Along with Kapteyn, Seeliger was not only one of the founders and one of the most important of the statistical cosmologists, but he also developed within Germany a network of colleagues, mostly former students, who continued to contribute to the growth of this subdiscipline. We have already noted Schwarzschild's important reformulation of the fundamental equation of stellar statistics that was later used by Kapteyn and van Rhijn in their construction of the Kapteyn Universe. Kienle also continued to produce important studies in statistical astronomy.[55]

Many of these enormous gains in international cooperation were nearly destroyed by the consequences of the First World War. As a result, when the Astronomische Gesellschaft decided to meet in August 1921 in Potsdam, Germany, the meetings represented a sort of watershed, because this was the first time after the war that scientists from allied, German, and neutral countries

---

[53] H. Shapley to H. Kienle, 11 October 1922 (Shapley).

[54] For example, writing to the astronomer K. H. G. Müller, Seeliger inquired about his sense of the political climate in Europe in order to hold the next meeting (in 1915) of the Astronomische Gesellschaft in St. Petersburg; H. v. Seeliger to K. H. G. Müller, 29 July 1914 (S. B. M.).

[55] Kienle, "Seeliger Festschrift"; also see Heilbron, *The Dilemmas of an Upright Man*, pp. 102 ff.

all participated. At the Potsdam meetings, a small group of colleagues from Denmark, Sweden, England, Germany, and Holland, including Kapteyn, met at Albert Einstein's house to discuss Einstein's General Theory of Relativity and his views on the structure of the stellar universe.[56] The only English representative at this meeting of the Gesellschaft, however, was Eddington, who believed it was absolutely essential to include the Germans. As Eddington wrote the Danish astronomer S. E. Strömgren, president-elect of the Gesellschaft at the time: "I hope to show my interest in the Astronomische Gesellschaft by attending the next meeting – an individual step which no one has any right to object to.... International Science is bound to win and recent events – the verification of Einstein's theory – has made a tremendous difference in the last month."[57] Though no evidence survives to confirm this view, it is most unlikely that Seeliger was present at this informal meeting, because he remained an uncompromising opponent of Einstein's theory of relativity.

INTERNATIONALIZING OF STATISTICAL COSMOLOGY

Following the emergence of the two major centers for statistical cosmology – in Groningen with Kapteyn, Pieter van Rhijn, and their Dutch students, and in Munich with Seeliger and his students, including most notably Karl Schwarzschild and Hans Kienle – beginning around 1912 in Sweden, Charlier directed his interests to many of the problems initially proposed by Seeliger's and Kapteyn's researches. Although Charlier's work was highly innovative mathematically, he and his later students generally followed the conceptual ideas suggested in the works of the Dutch and Germans, and Seeliger in particular, because he and Seeliger both shared a deep-seated mathematical propensity.

Following Kapteyn's lead, Charlier also made numerous international contacts. In 1916 he sent his student G. B. Strömberg to work with Hale as a stellar spectroscopist at Mount Wilson, where he remained until his retirement in 1946. A member of the

---

[56] A. S. Eddington, "Jacobus Cornelius Kapteyn, 1851–1922," *Royal Society of London, Proceedings* (Section A), 102 (1923), xxxv (xxix–xxxv).
[57] A. S. Eddington to S. E. Strömgren, November 1919, quoted in Hertzsprung-Kapteyn, *J. C. Kapteyn*, pp. 152–3.

Astronomische Gesellschaft since 1887, Charlier became a member of its Board of Directors in 1904 and served on the board until he resigned in 1917. He also became an honorary member of the American Astronomical Society, and received the Watson Gold Medal from the National Academy of Sciences. In 1933 he received the Bruce Medal from the Astronomical Society of the Pacific, as Kapteyn had earlier. During the spring and summer of 1924 Charlier spent time in California, lecturing, among other astronomical activities, at Berkeley.

As we have seen in Chapter 5, Eddington and Dyson also made notable contributions to the motions of stars and the "law" describing their space velocities. With the gradual acceptance of the theory of relativity, and after Eddington published in 1914 a masterly summary of (statistical) astronomy in his *Stellar Movements and the Structure of the Universe*, the leaders of astronomy in England plowed new and fertile ground.[58] After Bertil Lindblad and Jan Oort's work on galactic rotation between 1925 and 1927, which satisfactorily (and conclusively) explained Kapteyn's "two star-streaming" discovery, Eddington combined his older motion studies with his newly developing interests in relativity to deal with topics such as the rotation of the galaxy.[59]

With the singular exception of George Ellery Hale and his staff at Mount Wilson, the Americans, most notably Edward C. Pickering at Harvard and W. W. Campbell and Heber D. Curtis at the Lick Observatory, contributed, as did others, primarily to the stock of empirical data, though not as much to the construction of stellar models nor to their theoretical underpinnings. Thus, with few exceptions, Pickering and his staff produced invaluable photometric and photographic studies and stellar catalogues. And although the Lick astronomers engaged in leading work on stellar motions, by the early years of the twentieth century Seeliger, Kapteyn, and their research efforts focused more and more on the luminosity and density relationships, and not on the velocity function that had initially misled Kapteyn.

---

[58] Charlier's monograph, *The Motion and Distribution of the Stars* (Berkeley: University of California Press, 1926), is similar to Eddington's *Stellar Movements and the Structure of the Universe* (London: Macmillan & Co., 1914).

[59] A. S. Eddington, *The Rotation of the Galaxy* (Oxford: Clarendon Press, 1930).

Kapteyn, Seeliger, and to a lesser degree Charlier and Eddington not only focused on critical problems, developed research programs, and established unique methodological approaches to the solution of the sidereal problem, but they also founded or at least encouraged avenues for the dissemination of their research. Seeliger used mostly well-established German journals as his scientific outlets, primarily the international journal *Astronomische Nachrichten* for his shorter pieces and the various publications of the Academy of Sciences at Munich for his longer papers. The *Astronomische Nachrichten* was the premier international journal for many decades, but it is still possible that the international exposure of Seeliger's research was severely limited, because he chose to publish almost all of his most important results pertaining to the sidereal problem through the Munich Academy. Furthermore, Seeliger always wrote in German, employing a highly abstract, mathematical style that would not have been accessible to many observationally oriented astronomers. In private correspondence with Hans Kienle, Shapley confirmed this observation that "Seeliger's publications are not as accessible as some."[60] Shapley's view could mean one of several things: the availability of the journals (Munich Academy), the language (German), and/or the style (abstract mathematics). Whatever the case, it is a fact that neither Seeliger nor his students (with the exception of Schwarzschild) developed sustained contacts with English-speaking astronomers.

In contrast, not only did Kapteyn develop numerous international contacts, including the critical network with Hale and the Mount Wilson Observatory, but he also published virtually everything in the English language, studiously avoiding a dense and abstract scientific style. Moreover, Kapteyn founded the *Publications of the Astronomical Laboratory of Groningen*, an astronomical journal dedicated to publishing, particularly in the early years, scientific reports disseminating much of Kapteyn's own earlier important work. As his reputation grew and his international contacts expanded, Kapteyn turned increasingly to American astronomical journals, principally the *Astronomical Journal*, *Astrophysical Journal*, and the *Contributions of the Mount Wilson*

---

[60] H. Shapley to H. Kienle, 11 October 1922 (Shapley).

Observatory, the latter two of which Hale both edited and published from his Pasadena offices.

Charlier's considerable research over the course of nearly thirty years was published mostly in *Lunds Astronomiska Observatorium Meddelanden*, two series of observatory publications that he started shortly after arriving at Lund. Charlier and his assistants and students published 150 numbers filling ten volumes in these series. Because the vast majority of the research published in his *Meddelanden* deals directly one way or the other with statistical astronomy, it is clear that Charlier considered research in the space distributions and the true motions of stars to be a key problem area of contemporary astronomy.

In contrast to Seeliger's published work, the vast majority of scientific papers in both Kapteyn's and Charlier's scientific journals were written in the English language, which was rapidly becoming the lingua franca of the scientific community. Consequently, their work and that of their students was readily accessible to the growing international community of astronomers.

Perhaps the central feature characteristic of statistical cosmology was the clear focus on developing *models* of the stellar system. In contrast, the rise of astrophysics, beginning in the latter half of the nineteenth century, emphasized the special importance of spectroscopy in understanding the underlying *relationships* of stellar phenomena. This perhaps crucial difference was symbolically represented by the fact that while Kapteyn became the first research associate at Mt. Wilson, Henry Norris Russell (1877–1957), a specialist in spectroscopy and Shapley's mentor at Princeton, became the second research associate in 1918.

Even though many American astronomers played a distinctly central role in the rise of astronomy as an international science, one can easily be seduced into believing this represents the entire story. After all, during the earliest years of the twentieth century, the language of science was becoming less German and more English (that is, American English!). This is certainly reflected in Kapteyn's and Charlier's publications. With American financial backing, astronomy was increasingly being dominated by the big American telescopes, particularly at Mount Wilson – the 60" in 1908, the 100" in 1917, and eventually (in 1946) the 200" Palomar Telescope. Although there may be much to persuade one of this view after, say, 1920, during the period covered by this study, roughly 1890 to 1924, the Europeans played an equally large role in the internationalization of astronomy as the nexus of astronomy eventually crossed the Atlantic.

Part III

STATISTICAL COSMOLOGY AND THE SECOND ASTRONOMICAL REVOLUTION

8

THE DECLINE OF A RESEARCH PROGRAM

Discussions of the size and nature of the galactic system during the 1920s saw the emergence of a new and radically different cosmological perspective, in opposition to the classical sidereal models developed principally by Kapteyn and Seeliger. The newer views initiated by Harlow Shapley (1885–1972) of Mount Wilson eventually led to the decline of the statistical models of the Galaxy, such as the Kapteyn Universe, and to their total demise by 1930.

Histories of astronomy are in general agreement that three major developments led to the abandonment of the traditional theories of galactic structure: (1) Shapley's results on the spatial distribution of the globular clusters (1918); (2) the Lindblad–Oort theory of differential galactic rotation (1925–7); and (3) Trumpler's verification of interstellar light absorption (1930). This era, roughly from 1918 to 1930, was herculean in the history of astronomy. It has been variously called the "second great astronomical revolution," the "Shapley revolution," and the "golden days."[1] It was a period rich in ideas, and radical in its consequences.

With the emergence of these newer, radical views, on the one hand, and the traditional views of the statistical cosmologists, on the other hand, came a period of intense conflict. Formally, the disagreement reached its culmination on 26 April 1920 in a discussion between Harlow Shapley of the Mount Wilson Observatory (however his appointment to the directorship of the Harvard College Observatory was well under way at this time)

---

[1] See, respectively, Otto Struve, *The Universe* (Cambridge, Mass.: MIT Press, 1964), p. 157: Bart Bok, "Harlow Shapley and the Discovery of the Center of Our Galaxy," in Jerzy Neyman, ed., *The Heritage of Copernicus* (Cambridge, Mass.: MIT Press, 1974), p. 26; and Bertil Lindblad to Peter van de Kamp, quoted in Peter van de Kamp, "The Galactocentric Revolution, a Reminiscent Narrative," *Astronomical Society of the Pacific, Publications*, 77 (1965), 325 (325–35).

and Heber D. Curtis (1872–1942) of the Lick Observatory in what has since become known as the "Great Debate." This encounter took place at the National Academy of Sciences in Washington, at a meeting on "the scale of the universe."[2] In recent years, the nature and historical significance of this debate has been considered in great detail.[3] What has emerged is an understanding of the ongoing discussion between the protagonists in both the published and unpublished scientific literature, which is what motivated this clash in cosmological views in the first place. Scholars have considered in detail (and what I shall review briefly below) the ideas and empirical data that Shapley and others used to martial support for their radical cosmology. For our purposes, it is most important to assess the traditional cosmological views entailed in the ideas of the statistical astronomers that we have been considering so that we can understand the gradual dominance of one cosmological viewpoint and the eventual demise of another. The period of most intense discussion on the actual scale of the Universe began a few years before the Great Debate, which most clearly focused the conflict, and lasted to about 1922. Although a great deal remained to be explained before most astronomers could become totally convinced about the newer cosmology, by late 1922 the traditional (statistical) cosmology lay in ruins. Hence

[2] Harlow Shapley and Heber D. Curtis, "The Scale of the Universe," *National Research Council, Bulletin*, 2 (1921), 171–217. At the time, neither Shapley nor Curtis considered this a debate. See Bok, "Harlow Shapley," pp. 53–5, and C. Whitney, *The Discovery of Our Galaxy* (New York: Alfred Knopf, 1971), pp. 214–15. Otto Struve first gave this symposium its rubric "the Great Debate"; see Otto Struve, "A Historic Debate about the Universe," *Sky and Telescope*, 19 (1960), 398 (398–401), and Otto Struve and V. Zebergs, *Astronomy of the Twentieth Century* (New York: Macmillan Co., 1962), p. 416. As early as 1935, however, Edwin Hubble referred to this symposium as a "quasi-debate"; see Hubble, *The Realm of the Nebulae* (New Haven: Yale University Press, 1935), 87–8.

[3] See M. A. Hoskin, "The 'Great Debate': What Really Happened," *Journal for the History of Astronomy*, 8 (1976), 169–82, reprinted in idem, *Stellar Astronomy: Historical Studies* (Chalfont St. Giles, 1982), pp. 175–88; Robert W. Smith, *The Expanding Universe: Astronomy's 'Great Debate,' 1900–1931* (Cambridge: Cambridge University Press, 1982), pp. 77–90; Daniel Seeley and Richard Berendzen, "Astronomy's Great Debate," *Mercury*, 7 (1976), 67–71, 88; and M. A. Hoskin, "Shapley's Debate," in J. E. Grindlay and A. G. Davis Phillip, eds., *The Harlow–Shapley Symposium on Globular Cluster Systems in Galaxies* (International Union Symposium 126) (Dordrecht: D. Reidel Publ., 1988), pp. 3–9.

we will examine the *reaction* to Shapley's newer views by those who stood most to lose, both methodologically and theoretically, with the acceptance of Shapley's cosmology, namely, Kapteyn, Seeliger, and the results of statistical cosmology in general.

SHAPLEY'S COSMOLOGY

The first substantive suggestion of the inadequacies of the classical sidereal models came from Shapley's investigations. Shapley did not propose his new galactic model initially, however, because of certain anomalous features unexplained by the generally accepted (statistical) models of the Galaxy. He crusaded for a radical restructuring of galactic theory only after he became convinced that Adriaan van Maanen's (erroneous) observations of internal motions in the spiral M101 were correct and after his own continued discovery of cepheid variable stars in certain globular clusters.[4] Shapley's principal work deals with the discovery of the true center of our Galaxy and with the claim that the approximate size of the Milky Way should be increased by a factor of ten over the statistical values.[5] These developments emerged from a detailed investigation of clusters of stars, particularly globular clusters, during his employment at Mount Wilson from 1914 to 1921. In the first article of a long series of now classical papers begun in 1915, Shapley explained his concern with stellar clusters in general on the grounds that they were independent systems capable of yielding a greater understanding of our own galactic system:

> The object in taking up the investigations of . . . stellar clusters is twofold. First, it is hoped that considerable advance can be made in our understanding of the internal arrangement and physical characteristics of these objects. Secondly, and probably of more importance,

[4] A brief overview of this conflict is given in Richard Berendzen, Richard Hart, and Daniel Seeley, *Man Discovers the Galaxies* (New York: Science History Publications, 1976), part I.
[5] For a general discussion of Shapley's discovery of the galactic center, see Bok, "Harlow Shapley," 26–61. With the discovery of interstellar light absorption in 1930 by Robert Trumpler, Shapley's value for the size of the Milky Way was reduced by a factor of three.

Figure 21   Harlow Shapley, ca. 1920. (Courtesy of the Harvard University Archives)

it is hoped that through the acquisition of a clearer appreciation of stellar clusters, particularly of globular systems, some contribution can be made to the knowledge of our own galactic system. For it is quite obvious that a globular cluster, whether considered as entirely extraneous to our galactic domain or not, it is in itself a stellar system on a great scale – a stellar unit which without doubt must be comparable to our own galactic system in many ways, though differing from it fundamentally in others.[6]

In this paper, entitled "The General Problem of Clusters," Shapley outlined an extensive research program dealing with clusters. Two of the problems of particular interest here were the determination of distances of clusters and the relation of globular clusters to the galaxy as a whole.[7]

Shapley's interest in these two closely related problems can already be discerned in his earliest astronomical researches. In his dissertation, written under Henry Norris Russell at Princeton, he concentrated on eclipsing binaries as a tool to obtain stellar distances in excess of those obtainable by triangulation parallax methods. About the time that Shapley was studying eclipsing binaries, Henrietta Swan Leavitt at the Harvard College Observatory discovered that in the Magellanic Clouds the longest periods of the stars known as cepheid variables corresponded to those of the greatest brightness.[8] Four years later, in 1912, Leavitt formulated a relationship that led to a most remarkable conclusion: The brightnesses of the cepheids increased directly with their period.[9] When generalized, this empirical law, known as the period-luminosity relation, was almost immediately recognized (first by

[6] H. Shapley, "First Paper: The General Problem of Clusters," *Mount Wilson Observatory, Contributions*, no. 115 (1915), 213 (201–21).

[7] Shapley, "First Paper," 213–14. At the time, the distance problem was considered one of the two most significant problems in astronomy. The other concerned the nature of stellar evolution; see, for instance, Robert G. Aitken, "Recent Progress in Our Knowledge of the Universe," *Science*, 58 (1915), 381–7.

[8] H. S. Leavitt, "1777 Variables in the Magellanic Clouds," *Harvard College Observatory, Annals*, 60 (1908), 106–7.

[9] H. S. Leavitt, "Periods of 25 Variable Stars in the Small Magellanic Cloud," *Harvard College Observations, Contributions*, no. 173 (1908), 1–3.

Ejnar Hertzsprung) as a crucial yardstick for measuring the enormous distances to stellar clusters.[10]

As he was completing his Ph.D. at Princeton in 1913, Shapley inquired about job prospects with Frederick H. Seares, who had been Shapley's mentor when he was an undergraduate (at the University of Missouri) and who had subsequently taken a position at the Mount Wilson Observatory. Seares arranged for an interview with George Ellery Hale from which Shapley obtained an appointment as an assistant at Mount Wilson.[11] In transit to Pasadena, Shapley stopped at Harvard, where Leavitt had been doing her work. Pickering's assistant Solon I. Bailey suggested to Shapley during his stay that "when you get there, why don't you use the big [sixty-inch] telescope to make measures of stars in globular clusters?"[12] In fact, the first task Hale assigned to Shapley at Mount Wilson was to determine the cause of cepheid variation. These variable stars, as Bailey reminded Shapley, were particularly abundant in globular star clusters.[13] In the process Shapley began to examine cepheid variables in globular clusters as a means of extending Hertzsprung's work on stellar distances.

At first his investigations on cepheids in globular clusters led him to a theory of the structure and dimensions of the sidereal system not unlike that of Kapteyn and Seeliger. In his second paper of the series (1915), using Kapteyn's luminosity curve and a method developed by Hertzsprung (1913), he obtained distances to the globular clusters that suggested they were "complete and separate" systems, not a part of the galactic system.[14]

[10] E. Hertzsprung, "Ueber die räumliche Verteilung der Veränderlichen vom δ Cephei Typus," *Astronomische Nachrichten*, 196 (1913), 201–10. Cf. J. H. Jeans, "The New Outlook in Cosmogony," *Smithsonian Institution, Annals*, (1927), 151–60.

[11] H. Shapley to G. E. Hale, 14 November 1912; G. E. Hale to H. Shapley, 26 December 1912 (Hale). Shapley's position at the Mount Wilson Observatory, however, did not begin until 1 January 1914.

[12] H. Shapley, *Through Rugged Ways to the Stars* (New York: Charles Scribner's & Sons, 1969), p. 41.

[13] See B. J. Bok, "The Universe Today," in D. W. Corson, ed., *Man's Place in the Universe: Changing Concepts* (Tucson: University of Arizona Press, 1977), p. 100. We now refer to these variable stars – after their brightest prototype – as RR Lyrae stars.

[14] H. Shapley, "Second Paper: 1300 Stars in the Hercules Cluster (M13)," *Mount Wilson Observatory, Contributions*, no. 16 (1915), 307–8 (223–314).

Moreover, Shapley agreed with the statistical astronomers that a demonstration "that the maximum radius of the Milky Way is probably not greater than ten thousand light years [3,000 parsecs] and may be somewhat less has been deduced from many lines of evidence."[15] Thus the clusters and the galactic system appeared to be comparable in both size and form.

These views, especially with regard to the extent and form of the stellar system, matched the general views of astronomers on the nature of the Universe.[16] They also placed Shapley clearly on the side of the "island universe" proponents, who had been gaining adherents since about 1900 as a result of their continued examinations of spiral nebulae.[17] Shapley's "attack on star clusters" pleased Kapteyn personally, because it provided further support for the traditional model of a limited galactic system using techniques developed by Kapteyn and other statistical cosmologists.[18] Writing to Kapteyn in February of 1917 about this work, Shapley noted that "the work on clusters goes on monotonously – monotonous so far as labor is concerned, but the results are continual pleasure. Give me time enough and I shall get something out of the problem yet."[19] Within a year, however, Shapley had completely reversed his position both with respect to the size of the galactic system and to the island universe question. As he wrote to Eddington in January of 1918 on these newly altered and quite radical views:

> I have had in mind from the first that results more important to the problem of the galactic system than to any other question might be contributed by the cluster studies. Now, with startling suddenness and definiteness,

[15] Shapley, "Second Paper," p. 308.
[16] For a discussion of the contemporary conceptions of the sidereal universe at the time Shapley wrote his paper, see John S. Plaskett, "The Sidereal Universe," *Journal of the Royal Astronomical Society of Canada*, 9 (1915), 37–56.
[17] For the best contemporary discussion of the island universe controversy, see Smith, *The Expanding Universe*, chap. 1. For a review discussion of the growth of the island universe controversy based on the secondary published literature, see J. D. Fernie, "The Historical Quest for the Nature of the Spiral Nebulae," *Astronomical Society of the Pacific, Publications*, 82 (1970), 1189–230. Also, for period literature, see note 34.
[18] G. E. Hale to J. C. Kapteyn, 12 April 1915 (Hale).
[19] H. Shapley to J. C. Kapteyn, 6 February 1917 (Shapley).

they seem to have elucidated the whole sidereal system. . . .

To be brief, the globular clusters outline the sidereal system, but they avoid the plane of the Milky Way. . . . All of our naked-eyed stars, the irregular nebulae, eclipsing binaries – everything we know about, in fact, and call remote, except those compactly formed globular clusters, a few outlying cluster-type variables, the Magellanic clouds, and perhaps, the spiral nebulae. The globular clusters apparently can form and exist only in the parts of the universe where the star material is less dense and the gravitational forces less powerful than along the galactic plane. This view of the general system, I am afraid, will necessitate alterations in our ideas of star distribution and density in the galactic system.[20]

In 1918 he published three major papers in which he outlined a radically new arrangement of the sidereal universe.[21]

In the first of these papers Shapley showed that cepheids in the neighborhood of the Sun obeyed the period–luminosity relationship. Invoking the principle of the uniformity of nature, Shapley went on to assume that cepheids of a given period are comparable wherever found, whether near the Sun or in globular clusters.[22] The period–luminosity relation would therefore yield the absolute magnitude of the variables. Measuring the mean apparent magnitude and using the magnitude–distance relationship,[23] Shapley obtained distances to the clusters assuming, of course, negligible interstellar light absorption, a hypothesis for which he had secured some evidence. In the sequel submitted the following month,

[20] H. Shapley to A. S. Eddington, 8 January 1918 (Shapley).
[21] For complementary accounts of the development of Shapley's galactic model, see Smith, The *Expanding Universe*, pp. 59–77, and Owen Gingerich, "Shapley, Harlow," in C. C. Gillispie, ed., *Dictionary of Scientific Biography* (New York: Charles Scribner's & Sons, 1975, 16 vols.), vol. 12, pp. 346–7 (345–52).
[22] H. Shapley, "Sixth Paper: On the Determination of the Distances of Globular Clusters," *Mount Wilson Observatory, Contributions*, no. 151 (1917), 115–16 (81–116).
[23] The magnitude–distance relationship relates the absolute magnitude (M) of a star with its apparent magnitude (m) and its distance (r) in parsecs. Thus, $M = m + 5 - 5 \log (r)$.

Shapley performed the actual calculations discussed in the earlier paper. His results were dramatic, contradicting all contemporary conceptions of the cosmos. The distances of the clusters, sixty-nine all together, ranged from 6,000 to 67,000 parsecs.[24] Plotting the positions of the clusters onto a plane through the Sun and perpendicular to the galactic plane, he obtained a figure that gave the distribution and center of the system of clusters. The analysis suggested that the sixty-nine clusters defined a system whose central region was about 15,000 parsecs (50,000 light years) from the Sun in the direction of the Sagittarius star cloud.[25] In his third paper, which has since become the most famous of the many papers he published under the Mount Wilson Observatory rubric, Shapley described in great detail his radically unorthodox cosmology.[26] Building on his earlier results, he outlined a sidereal universe whose center coincided with that of the globular clusters, thus suggesting a galactic system about 90,000 parsecs (300,000 light years) in diameter. This necessitated removal of the Sun to a region some 15,000 parsecs from the galactic center. As for the island universe theory, he rejected it entirely: "The adoption of such an arrangement of sidereal objects leaves us with no evidence of a plurality of stellar 'universes'."[27]

At the time it was generally agreed that the Sun was near the center of the stellar universe and that the entire system of stars

[24] Although cepheids eventually became a major tool in distance determination, Shapley used cepheids in only about 10% of the calculated distances to the clusters, because not all clusters contained easily measured variables. In addition he used other methods based on the apparent magnitudes of the brightest stars in clusters and on their angular diameter. See H. Shapley, "Seventh Paper: The Distances, Distribution in Space, and Dimensions of 69 Globular Clusters," Mount Wilson Observatory, Contributions, no. 152 (1917), 117–31 (117–44).

[25] Shapley, "Seventh Paper," pp. 131–40. Only one astronomer before Shapley mentioned this fact. In 1909, the Swedish astronomer Karl Bohlin pointed out the asymmetrical distribution of the clusters in the hemisphere, suggesting that the center of the galaxy lay in the direction of the Sagittarius star cloud; see K. Bohlin, "On the Galactic System with regard to its Structure, Origin, and Relations in Space," *Kungliga Svenska Vetenkapsakademiens Handlingar*, no. 10 (1909), 43–65.

[26] H. Shapley, "Twelfth Paper: Remarks on the Arrangement of the Sidereal Universe," *Mount Wilson Observatory, Contributions*, no. 157 (1918), 209–34.

[27] Shapley, "Twelfth Paper," p. 209.

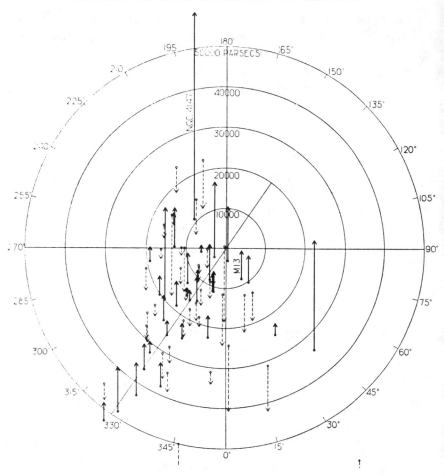

Figure 22 Distribution in space of globular clusters. The Sun is at the center of the diagram, and the globular clusters are maximally distributed along galactic longitude 325° in the direction of Sagittarius. (From "The Distances, Distribution in Space, and Dimensions of 69 Globular Clusters," *Mt. Wilson Observatory, Contributions*, no. 152 (1971), pp. 117–31)

was somewhere between 9,000 and 18,000 parsecs in diameter. Thus there was no corroborative evidence for Shapley's views. What then led Shapley to assert the centrality of the globular clusters as defining the galactic center?[28] A number of important factors provided evidence for Shapley's highly unorthodox position.[29] First, by 1918 the Sagittarius star cloud was considered unusually rich and dense in stellar objects, suggesting to others besides Shapley that the galactic center lay in this direction.[30] Second, of the sixty-nine clusters, all but four lie in the general direction of Sagittarius, suggesting the asymmetry of the Solar System with respect to the Milky Way. Third, the equatorial plane of the globular clusters appeared to be identical to the galactic plane. And fourth, Shapley's latest researches on globular clusters had suggested sizes for the clusters that were much smaller than even the currently accepted dimensions of the classical sidereal system; hence, Shapley could reasonably argue for their dependence upon the larger system of stars.[31]

Controversy and debate ensued after the appearance of Shapley's papers outlining his galactic structure of the sidereal universe.[32] The evidence of the eccentric distribution of the globular clusters was too overwhelming to be denied in light of the fact that they were considerably smaller than even the classical models of the sidereal system. It was not a question, therefore, of the peripheral location of the Sun that disturbed astronomers; rather it was the

---

[28] Most accounts have failed to treat this problem adequately; see Whitney, *The Discovery of Our Galaxy*, p. 212, and Bok, "Harlow Shapley," pp. 44–52.

[29] Bart Bok, "Harlow Shapley – Cosmographer and Humanitarian," *Sky and Telescope*, 44 (1972), 354–7.

[30] See Walter Baade, *Evolution of Stars and Galaxies*, ed. Cecilia Payne-Gaposchkin (Cambridge, Mass.: Harvard University Press, 1963), p. 9, and C. D. Perrine, "The Nature of Globular Clusters," *The Observatory*, 40 (1917), 166–8.

[31] Moreover, the groupings of the clusters around the galactic poles suggested to Shapley their dependence upon the larger system of stars. See Shapley, "Seventh Paper," pp. 136–40; Shapley, "Twelfth Paper," pp. 218–22; and H. Shapley, "Note on the Explanation of the Absense of Globular Clusters from the Mid-Galactic Regions," *Observatory*, 42 (1919), 82–4.

[32] Shapley himself wasted no time in publishing his ideas and conclusions in other technical journals to spread the new dogma; see H. Shapley, "Globular Clusters and the Structure of the Galactic System," *Astronomical Society of the Pacific, Publications*, 30 (1918), 42–54.

Figure 23   Shapley's asymmetric universe (1918).

enormous size of Shapley's stellar universe, implied in his claim
that the center of the system of globular clusters was identical
with the center of the stellar system, which, on the one hand,
militated against the island universe theory, and, on the other
hand, vitiated the implications of the classical statistical models
of the stellar system. Although it is not our purpose to discuss the
status of the island universe concept during this period, nearly all
advocates of the island universe theory fully accepted the general
conclusions of the statistical models.[33] In fact, aside from Shapley's
highly unorthodox approach to studying the stellar distribution,
contemporary attempts at stellar understanding of the sidereal
problem generally utilized the methods that statistical astronomy

[33]   Although it has been claimed that Curtis accepted the essential features of
the Kapteyn Universe, Curtis was impressed more by Kapteyn's methods
than by Kapteyn's conclusions. See Berendzen, Hart, and Seeley, *Man
Discovers the Galaxies*, p. 39. Curtis himself made it clear that "while I am
ready to worship Kapteyn's *methods*, in which he has been fifty years ahead
of the times, I can not, as most astronomers do, fall down and worship all
the results which have come out of this mathematical mill." H. D. Curtis to
Alter, 1 February 1922 (Allegheny Observatory Archives). I am indebted to
Robert W. Smith for bringing this letter to my attention.

had developed over many years. Thus when Shapley rejected the island universe theory, which was favored by most of his contemporaries, he also rejected the leading theories of galactic structure.

At the time, the technical and semipopular literature was replete with discussions of the island universe theory and, in particular, the status of the spiral nebulae.[34] The majority of astronomers favored the island universe theory; even Shapley was orthodox on this issue as late as 1915, and not until 1917 did he actively oppose the island universe theory. Within a year, only Shapley and a few others advocated the newer position. Shapley, however, had not examined the spirals for variable stars. His evidence for the close proximity of the nebulae was obtained from two other sources: (1) Adriaan van Maanen, Shapley's colleague at Mount Wilson, thought he had detected internal motions in a spiral, which, if an external galaxy, implied velocities in excess of the speed of light;[35] and (2) a surprisingly bright nova

---

[34] For some of the best period accounts of the development and controversy surrounding the island universe theory, see the following: F. W. Very, "Are the White Nebulae Galaxies?" *Astronomische Nachrichten*, 189 (1911), 441–54; P. Puiseaux, "Spiral Nebulae," *Revue Scientifique*, April 6, 1912; A. C. D. Crommelin, "Are the Spiral Nebulae External Galaxies?" *Scientia*, 21 (1917), 365–76; H. D. Curtis, "Modern Theories of Spiral Nebulae," *Journal of the Washington Academy of Sciences*, 9 (1919), 217–27; H. Shapley, "On the Existence of External Galaxies," *Astronomical Society of the Pacific, Publications*, 31 (1919), 261–8; H. MacPherson, "The Problem of Island Universe," *Observatory*, 42 (1919), 329–34; and D. B. McLaughlin, "The Present Position of the Island Universe Theory of the Spiral Nebulae," *Popular Astronomy*, 30 (1922), 286–95, 327–39.

[35] See A. van Maanen, "Preliminary Evidence of Internal Motion in the Spiral Nebula Messier 101," *Mount Wilson Observatory, Contributions*, no. 118 (1916), 331–49; and A. van Maanen, "Internal Motions in a Spiral Nebulae," *Observatory*, 39 (1916), 514–15. For some independent corroborative evidence on this problem, see J. H. Jeans, "Internal Motions in Spiral Nebulae," *Observatory*, 40 (1917), 60–1. Van Maanen's classic diagrams of the rotation of spirals appeared in "Investigations on Proper Motions. Fourth Paper: Internal Motion in the Spiral Nebula Messier 51," *Mount Wilson Observatory, Contributions*, no. 213 (1923), 221–9, and "Investigations on Proper Motions. Fifth Paper: Internal Motion in the Spiral Nebula Messier 81," *Mount Wilson Observatory, Contributions*, no. 214 (1923), 231–40. For a discussion of the larger context, see Norris Hetherington, "Adriaan van Maanen and Internal Motions in Spiral Nebulae: A Historical Review," *Quarterly Journal of the Royal Astronomical Society*, 13 (1972), 25–39; idem, "Adriaan van Maanen on the Significance of Internal Motions in Spiral

had appeared in the Andromeda Nebulae in 1885 that, under the assumption that it was comparable to novae within the Milky Way, suggested the nebulae must be within the larger sidereal system.[36] But other evidence from the spirals mitigated Shapley's argument, particularly the unusually high radial velocities with negligible proper motion detected by V. M. Slipher, as well as the detection of other novae already by 1917.[37] Slipher's results implied, under the reasonable assumption of transverse motions comparable to radial velocities, enormous distances of up to 6 million parsecs (20 million light years). The conflicting positions on the island universe theory and galactic dimensions eventually led to the Great Debate between Shapley and Curtis. Curtis, in concert with his director, W. W. Campbell, had been Shapley's chief public opponent on the island universe question ever since Shapley announced his radical cosmology in 1918.

THE DUTCH REACTION

The question of the distances and general dimensions of Shapley's stellar system was another matter. Concerning Shapley's model of the Galaxy, the German-American astronomer Walter Baade has written:

> I have always admired the way in which Shapley finished this whole problem in a very short time, ending up with a picture of the Galaxy that just about smashed up all the old school's ideas about galactic dimensions. . . . It was a very exciting time, for these distances

---

Nebulae," *Journal for the History of Astronomy*, 5(1) (1974), 57–8; idem "The Simultaneous 'Discovery' of Internal Motions in Spiral Nebulae," *Journal for the History of Astronomy*, 6(2) (1975), 115–25; idem "Edwin Hubble on Adriaan van Maanen's Internal Motions in Spiral Nebulae," *Isis*, 65 (1974), 390–3; and idem, *Science and Objectivity* (Ames: Iowa State University Press, 1988).

36 See A. C. D. Crommelin, "The Nature of Spiral Nebulae," *Journal of the British Astronomical Association*, 28 (1918), 177–9.

37 V. M. Slipher, "Spectroscopic Observations of Star Clusters," *Popular Astronomy*, xxvi (1918), 8.

seemed to be fantastically large, and the "old boys" did not take them sitting down.[38]

The technical literature clearly substantiates Baade's observation that the "old boys" strenuously objected to Shapley's radical views on this matter. As regards the distance problem introduced by Shapley's model, the old boys turn out to be Kapteyn and the Dutch school of statistical astronomers.

As we have seen, around 1900 two major centers of statistical cosmology, one in Holland under Kapteyn's astute leadership and the other in Germany under the direction of Seeliger, began to flourish. Although significant contributions to statistical astronomy were later made by Charlier, and, to a lesser extent, by Eddington and Dyson, both the Swedish and English astronomers, and surprisingly even Seeliger, were not nearly as insistent as Kapteyn and his Dutch colleagues were in rejecting Shapley's claims. Others also objected to Shapley's cosmology, but they did so mostly for its implications for the island universe theory.

Although his 1920 system was nearly Sun centered, Kapteyn was not willing to relinquish the heliocentric assertion readily, despite increasingly stronger evidence from others' research, principally from Shapley. Kapteyn had earlier written Shapley that because the Milky Way appeared to be similar in all directions, he could not accept Shapley's assertion of an eccentric position for the Sun.[39] It was not only Shapley's rejection of the idea of a Sun-centered system that disturbed Kapteyn, but also Shapley's challenge to the reality of Kapteyn's luminosity curves: "I was much interested in this 'second approximation' [1920]. There are many points I would be glad to discuss with you, particularly the matter of symmetrical luminosity curves. I cannot convince myself that they exist except as the reflection of the combination of many diverse factors."[40] Because he was one of the leading proponents of the traditional statistical approach to understanding galactic structure, Kapteyn's life's work stood in jeopardy,

---

[38] Baade, *Evolution of Stars and Galaxies*, p. 9. Shapley himself referred to the conservative old boys as "patriarchs"; see Hoskin, "The 'Great Debate'," p. 172.

[39] J. C. Kapteyn to H. Shapley, 15 June 1919 (Shapley).

[40] H. Shapley to J. C. Kapteyn, 7 March 1920 (Shapley).

particularly his conception of the stellar system, which was based on his technique utilizing stellar luminosities.

Thus both Shapley's methods and his results presented a major challenge to the established views. Even so, Kapteyn's relationship with Hale and the Mount Wilson Solar Observatory staff (which at the time included Shapley) was highly productive and they generally remained on excellent terms.[41] As we have seen, Kapteyn's work at Mount Wilson began in 1908 when he became a research associate, spending each summer on the mountain until the outbreak of the First World War. Those years were highly productive and rewarding for the Mount Wilson people. "Nothing has been so valuable to the Mount Wilson Observatory," wrote its director Hale in 1922, "as the inspiration and guidance of Kapteyn. His splendid imagination, great breadth of view and fine optimism have stimulated us to our best efforts and encouraged us to attack problems of wide scope."[42] Despite what he understood as the life-and-death conflict between two radically opposing theories, Kapteyn carried on his prolonged disagreement with Shapley, but with great respect for Shapley's abilities. When Edward C. Pickering died on 3 February 1919, Kapteyn strongly urged Hale to lobby for a statistical astronomer as Pickering's replacement as director of the Harvard University Observatory, because "I acutely feel that the succession in Harvard may thoroughly affect the development of statistical astronomy in the near future."[43] However prophetic these words were to become in the context of the two men's fundamental differences over stellar structure, it is revealing that Kapteyn endorsed

---

[41] On only two occasions was Kapteyn's relationship with the Mount Wilson people marred: Once over a professional dispute with Walter S. Adams in 1917, in which Kapteyn resigned (temporarily) his formal relationship with the Mount Wilson Observatory, and later over Hale's disagreement with Kapteyn concerning the role of German scientists in the First World War. See, respectively, W. S. Adams to J. C. Kapteyn, 24 April 1917; J. C. Kapteyn to G. E. Hale, 26 June 1917; G. E. Hale to J. C. Kapteyn, 22 September 1917; J. C. Kapteyn to G. E. Hale, 18 November 1917, 3 December 1917, 31 August 1919; and G. E. Hale to W. de Sitter, 10 February 1921 (Hale).

[42] G. E. Hale to W. de Sitter, quoted in W. de Sitter, "Nekrologe – Jacobus Cornelius Kapteyn," *Vierteljahrsschrift der Astronomische Gesellschaft*, 58 (1923), 182 (162–90).

[43] J. C. Kapteyn to G. E. Hale, 7 February 1919 (Hale).

Shapley's appointment (in 1921) as director at Harvard: "You know that I have always thought of him [Shapley] as one of the best, or the best man for the position."[44] Despite some minor concessions to Shapley, Kapteyn's fundamental disagreement over the new cosmological views remained intractable.

The Dutch School's first sustained attack on Shapley's distances came from Kapteyn's last doctoral student, Willem J. A. Schouten, who finished a dissertation on the methods of statistical astronomy in 1918, the year in which Shapley first published his new views.[45] Early in 1918, before Shapley had published the details of his radically new model but after he had already determined the enormous distances to the globular clusters, Schouten explicitly challenged Shapley's conclusions.[46] Using a method based on Kapteyn's luminosity-curve that was fully outlined by Kapteyn in 1914,[47] Schouten obtained (statistical) parallaxes to about a dozen globular clusters, including some of those for which Shapley had derived values. Examining Messier 13, Schouten found a parallax value of 0.00075, whereas Shapley had earlier obtained a value of 0.00008. Overall, Schouten's numbers suggested that the clusters were on the average seven to eight times closer than as represented in Shapley's work. Consequently, following Shapley's assumption that the globular clusters defined our stellar system, Schouten's analysis revealed a stellar universe about 12,300 parsecs

---

[44] J. C. Kapteyn to G. E. Hale, 3 January 1922 (Hale).

[45] The influence that Kapteyn exerted on his pupils, as well as on others, was enormous. Both van Rhijn and Schouten, who completed their dissertations under Kapteyn in 1915 and 1918, respectively, were among the old boys referred to by Baade. Although they were roughly Shapley's age, both were in the forefront in defending Kapteyn's lifelong project to solve the sidereal problem directly by the use of classical statistical methodology.

[46] W. J. A. Schouten, "On the Parallax of some Stellar Clusters," *Koninklijke Akademie van Wetenschappen te Amsterdam. Proceedings of the section on sciences*, 20 (1918), 1108–18 (communicated 26 January 1918); W. Schouten, "On the Parallax of Some Stellar Clusters," *Koninklijke Akademie van Wetenschappen te Amsterdam. Proceedings of the section on sciences*, 21 (1919), 36–47 (communicated 23 February 1918). Also see W. Schouten, "The Parallax of Some Stellar Clusters," *Observatory*, 42 (1919), 112–19 (communicated November 1918).

[47] See J. C. Kapteyn, "On the Individual Parallaxes of the Brighter Galactic Helium Stars in the Southern Hemisphere, Together with Considerations on the Parallax of Stars in General," *Mount Wilson Observatory, Contributions*, no. 82 (1914), 400–1 (339–442).

(40,000 light years) in diameter, almost precisely the size of the Kapteyn Universe.

Schouten's technique was based on Kapteyn's original luminosity-curve and on the supposition that the spread of stellar brightnesses was identical throughout space, whether within the local solar neighborhood or within globular clusters. Schouten felt justified in using the luminosity-curve method and making the uniformity claim of stellar brightnesses, because, in his judgment, the luminosity-curves of various clusters greatly resembled those Kapteyn found for stars in the solar neighborhood.[48] Given proper motion and apparent motion data, this technique indicated that, statistically speaking, high proper motion cepheid variables must be fairly close, and therefore these stars must be dwarfs, and not giant stars (in globular clusters) at large distances, as Shapley suggested.

Using traditional methods of statistical astronomy that had been in development since the latter years of the nineteenth century, others among the old boys also sought to discredit the unorthodox approach Shapley pioneered. Their efforts were not, however, entirely unproblematic. For instance, in 1919 the Dutch astronomer Antonie Pannekoek published a second method of parallax determination, which was also based on the luminosity-curve and on the supposition of the uniformity of physical law throughout the Universe.[49] The results of his method challenged those of Schouten and tended to confirm those of Shapley. Analyzing the Cygnus and Aquila star clouds, Pannekoek obtained distances of 40,000 and 60,000 parsecs, respectively. Earlier Shapley had obtained maximum values of 67,000 parsecs to various globular clusters. Because Pannekoek's work was based on the latest determination of Kapteyn's luminosity function, as evaluated by Schouten in his 1918 dissertation, the Dutch school did not take his work lightly. A few years later (1921), the semi-professional Dutch astronomer Cornelis Easton, who had spent much of his professional career studying the Aquila and Cygnus clouds, countered Pannekoek and argued that these distances

[48] Schouten, "On the Parallax of Some Stellar Clusters," 36–47.
[49] A. Pannekoek, "The Distance of the Milky Way," *Royal Astronomical Society, Monthly Notices*, 79 (1919), 500–7.

must be less than 10,000 parsecs.[50] Both Willem J. Luyten, who was among the second generation of Dutch statistical astronomers, and Kapteyn advised Easton in his analysis.

The central issue of stellar distances to clusters, spirals, and star-clouds was very much debated during the early 1920s. The Shapley–Curtis symposium on "the scale of the universe" did not resolve the distance problem or significantly alter the status of the island universe theory; at best it simply clarified assumptions required with either position. Fundamentally, both Shapley and Curtis agreed that the stellar system was, above all, uniform in its characteristics; that is, for instance, stars of a particular spectral type would exhibit basically the same absolute magnitude whether in our "local" system or in a cluster. Indeed, Shapley's cosmology was absolutely predicated on this supposition. As the German astronomer A. Kopff expressed it at the time: "With the principal of the physical homogeneity of the Cosmos stands or falls the work which Shapley has undertaken."[51]

What Kopff had in mind here was Shapley's crucial assertion that cepheid variables have uniform characteristics wherever they might be located. Shapley (and others) had shown that long-period cepheids are giant stars.[52] What Shapley had assumed, therefore, was that short-period variables in clusters are also giant stars analogous both to the short-period cepheid variables in the stellar system in general and to the long-period cepheids in clusters. Even though Shapley was aware that the long-period cepheids tend to be grouped near the galactic plane, whereas the short-period cepheids are more evenly distributed, he glossed over this distinction in his analysis of the cepheid variables.[53] It was precisely on this point that Kapteyn felt that Shapley's argument was in error. As early as March 1920, in a long letter to Hale, Kapteyn proposed several tests "meant for settling the conflict

[50] C. Easton, "On the Distance of the Galactic Star-Clouds," *Royal Astronomical Society, Monthly Notices*, 81 (1921), 215–26.
[51] A. Kopff, "Die Untersuchungen H. Shapleys über Sternhaufen und Milchstrassensystem," *Die Naturwissenschaften*, 9 (1921), 769 (769–74).
[52] Shapley, "Sixth Paper," pp. 81–116.
[53] See H. Shapley, "Eighth Paper: The Luminosities and Distances of 139 Cepheid Variables," *Mount Wilson Observatory, Contributions*, no. 153 (1917), 148–54, 160 (145–60).

between the parallaxes found for the globular clusters by Shapley and Schouten."[54] At the same time, Kapteyn also sent Shapley a précis of this longer letter, which contained an olive branch: "I hope to live in order to see your and my studies meet."[55] In one way or another, Kapteyn's proposals were predicated on the notion that short-period and long-period cepheids are in a fundamental way radically different from one another with respect to their distances. Precisely two years later, in March 1922, Kapteyn, in collaboration with Pieter van Rhijn, published the results of investigations of these proposals, attacking Shapley on the crucial point of the general uniformity of stellar distributions and of masses among cepheids. In contrast to Shapley's view, they emphasized the possibility that the cepheid variable stars in clusters are actually dwarfs, and not giants, indicating that the clusters are quite close. Restricting their analysis to the short-period cepheids, they obtained a mean-parallax value 7.6 times larger than Shapley's.[56] The implication, of course, was that Shapley's model was much too large and should be reduced by a factor of almost eight. Rather than being 90,000 parsecs (300,000 light years) in diameter, the stellar system, if one was to rely on Shapley's idea of utilizing cepheid variables, should be reduced to about 12,000 (90,000/7.6) parsecs the equivalent of about 40,000 light years – which, as Kapteyn pointed out rhetorically, "agrees exactly with Schouten's value."[57] This view fundamentally contradicted Shapley's claim, known to Kapteyn and van Rhijn since 1917, that in consequence of the simultaneous occurrence of both long-period and short-period cepheid variables in globular clusters, short-period cepheids must also be giant stars. As Shapley implied (and as we know today), high proper motion cepheids of short-period are giants, and because they possess high space velocities, they are nevertheless to be placed at great distances.[58]

[54] J. C. Kapteyn to G. E. Hale, 3 March 1920 (Hale).
[55] J. C. Kapteyn to H. Shapley, 4 March 1920 (Shapley).
[56] J. C. Kapteyn and P. van Rhijn, "The Proper Motions of Cephei Stars and the Distances of the Globular Clusters," *Bulletin of the Astronomical Institutes of the Netherlands*, i (1922), 40–1 (37–42).
[57] Kapteyn and van Rhijn, "The Proper Motions of Cephei Stars," p. 41.
[58] H. Shapley, "Notes Bearing on the Distances of Clusters," *Harvard College Observatory, Contributions*, no. 237 (1922), 9–10 (1–11).

Hoping for more time to reconsider these conflicting results, Kapteyn was willing to settle for a temporary truce on the question of the arrangement of the stellar universe: "The present paper will have fulfilled its aim if . . . astronomers suspend judgment on all the important deductions involving the knowledge of the parallax of the clusters."[59] Although Kapteyn was attempting to regain the initiative in this matter by attacking Shapley's fundamental assumptions directly, this and other statements suggest that he understood fully the strength of Shapley's position and, consequently, assumed a conservative, defensive posture.

The differences between Kapteyn and Shapley began in an earnest, but friendly and even accommodating manner. Here the older man, who had worked to understand the structure of the sidereal universe for over forty years, was being challenged by the much younger, though highly innovative American. Although they concerned themselves with the same general problem, their focuses and methods of analysis were quite different. In a letter sent directly to Shapley in June 1919, Kapteyn cryptically summarized their differences:

> Meanwhile, though at least in the last papers [of Shapley's cluster series] you treat of the general arrangement of the sidereal universe, the subject which, more than anything else, forms the subject of our own [Groningen] Laboratory's work, your work meets ours but very little. I have to repeat what, at the time, I said to [Cornelis] Easton. You are building from above, while we are up from below. You start from the general system in its greatest extension, we try to struggle laboriously up from our nearest [stellar] surroundings. When will the time come that we [our results] thoroughly meet? Heaven knows. That for me the chances are bad. I am now 68 and will leave the University and with it the Laboratory, in about 2 years. That our investigations will then have come together I dare not hope and what I can do afterwards there is no knowing. You will therefore not expect me to enter into any serious, somewhat general discussion of your work. I feel unable to do it.

[59] Kapteyn and van Rhijn, "The Proper Motions of Cephei Stars," p. 41.

Still as a proof of my great interest, I may perhaps make a few unconnected remarks.[60]

Kapteyn's "unconnected remarks" were later greatly expanded in his March 1920 letter to Hale, and were fully developed in his 1922 critique of Shapley's scheme.

By 1922, their differences, though still expressed in a cordial manner, had reached serious proportions. It was clear to the antagonists involved that there were irreconcilable differences that needed urgent attention if the future progress in sidereal studies were to continue in a healthy and productive atmosphere. During the later spring of 1922 an ideal opportunity arose that made it possible for Shapley to meet face-to-face with the Dutch astronomers. Shapley had already made plans to attend the International Astronomical Union in Rome, Italy. As a result, Kapteyn, Hertzsprung, van Rhijn, Schouten, and Pannekoek, in concert with Shapley, planned a special colloquium of the Dutch Astronomical Society that would be held after the Rome meetings to discuss the dimensions, extent, and structure of the galactic system.[61] Held in Leiden on May 27, this meeting, which Kapteyn did not attend due to a terminal illness that would take his life within a month's time, allowed Shapley to "argue with the experts there on the scale of the universe."[62] Although there were no recorded minutes of the discussion, personal correspondence reveals much of the substance of the conclusions. In their March 1922 critique of Shapley, Kapteyn and van Rhijn had apparently overlooked the fact that the short-period cepheids, which they used to demonstrate the nearness of globular clusters, mostly possessed high *radial* velocities, a fact that invalidated their method of calculating the parallactic motions of the stars.[63] Therefore, as Shapley expressed it at the time, "the big proper motion means big space velocity . . . and does not mean nearness, low luminosity, and parallaxes eight times larger for the globular

---

[60] J. C. Kapteyn to H. Shapley, 15 June 1919 (Shapley).
[61] H. Shapley to A. Pannekoek, 28 February 1922; H. Shapley to P. van Rhijn, 26 March 1922; H. Shapley to George R. Agassiz, 20 March 1922 (Shapley).
[62] H. Shapley to G. R. Agassiz, 19 June 1922 (Shapley).
[63] H. Shapley to Peter Doig, 21 June 1922; H. Shapley to Knut Lundmark, 15 July 1922 (Shapley).

clusters."[64] The parallaxes "eight times larger," of course, referred to Kapteyn's much smaller sidereal universe.

Not only was Shapley adamant about his interpretation of
the data, but some of the Dutch scientists themselves did not
take their own compatriots' work seriously, particularly that of
Schouten.[65] Even van Rhijn admitted personally to Shapley, while
the latter was still in Holland, that the conclusions he and Kapteyn
had reached were "hasty."[66] After the Leiden meeting, the trickle
became an avalanche, if we are to believe Shapley entirely. Besides van Rhijn and Hertzsprung, many others, including those
not directly involved in this major dispute, such as Bertil Lindblad,
Andrew Crommelin, Henry Norris Russell, M. Hopman, Walter
Baade, and Arnold Kohlschütter, agreed that the conclusions of
Kapteyn and van Rhijn were simply not justified.[67] Speaking before
the British Astronomical Association after the Dutch meeting,
Shapley, in no uncertain terms, again resumed his offensive thrust:

> The present discussion [on the galactic system] is made
> in the light of criticisms and numerous tests to which
> the conclusions have been subjected during the past
> four or five years. . . . [A] sufficient answer to those
> who would reduce the distance of clusters to one-fifth
> or one-tenth the values proposed, is that apparently
> they do not consider fully the dire consequences of
> such reduction on a vast body of other astronomical data
> that is generally accepted.[68]

A much reduced system would, in Shapley's opinion, negate the
fundamental supposition of the uniformity of space and physical
law, and thus jeopardize "a vast body of astronomical data."

In the last published paper of his career (1922), the substance
of which he presented in November 1921 in Leiden to an informal gathering of specialists that included Albert Einstein and

[64] H. Shapley to Isabel M. Lewis, 7 July 1922 (Shapley).
[65] H. Shapley to P. Doig, 21 June 1922 (Shapley).
[66] H. Shapley to I. M. Lewis, 7 July 1922; H. Shapley to H. Kienle, 27 September 1922 (Shapley).
[67] H. Shapley to K. Lundmark, 15 July 1922 (Shapley).
[68] A revised version of Shapley's BAA talk appeared as "The Galactic System,"
Nature, 110 (1922), 545–7, 578–81.

James Jeans, Kapteyn presented his final resume of ideas concerning the sidereal system in terms that, it may be inferred albeit erroneously, reflect an understanding of Shapley's work. Specifically, in marked contrast to his 1920 stellar model, he rejected the "erroneous assumption that the sun is the centre."[69] Instead, as we have already seen in Chapter 6, he argued that the galactic center was roughly 650 parsecs (2,300 light years) from the Sun. This was not a symbolic concession to Shapley, however! Kapteyn placed the Sun asymmetrically within his new system in order to provide a physical basis for the star-streaming phenomenon.

THE GERMAN AND SWEDISH RESPONSES

Aside from island universe proponents like Heber Curtis, those who favored the "old school" and sought to reaffirm its validity were statistical astronomers, and most of them were Dutch. The question remains why it was chiefly the Dutch who adamantly opposed Shapley in the technical literature, and not Seeliger and a host of other statistical astronomers who trained under this "dean" of early twentieth-century German astronomy.[70]

Germany had become isolated during the First World War, and so German scientists and intellectuals generally remained out of touch with current research and developments elsewhere in Europe and the United States. In fact, when Shapley's series of papers on stellar clusters appeared beginning in 1915 in the *Mount Wilson Solar Observatory Contributions*, the circulation department of the parent organization (the Carnegie Institution in Washington, which was responsible for distributing the *Contributions*) did not send these publications to Europe.[71] Among the few European astronomers who regularly received Shapley's papers on stellar clusters were Kapteyn and van Rhijn, and this was only because Shapley personally sent them to his Dutch

---

[69] J. C. Kapteyn, "First Attempt at a Theory of the Arrangement and Motion of the Sidereal System," *Astrophysical Journal*, 55 (1922), 321 (302–27).

[70] Otto Struve was an on-site observer at conferences in which Seeliger was perceived and treated as the "dean" of German astronomy; see Struve and Zebergs, *Astronomy of the Twentieth Century*, pp. 39–40.

[71] H. Shapley to G. E. Hale, 19 January 1918 (Hale).

colleagues.[72] In contrast, Seeliger never became sufficiently familiar with Shapley's work, primarily because American publications
rarely entered Germany during the war, and only slowly thereaf-
ter. (During the last years of the war, it was illegal for an American
to send mail to Germany.)

One of Seeliger's biographers, Dr. Felix Schmeidler of the Institut
für Astronomie und Astrophysik at the University of Munich,
suspects that Seeliger *never* became familiar with Shapley's work,
even after the war.[73] This view is supported by the fact that Hans
Kienle, Seeliger's former student and colleague at the Munich
Observatory both during and after the First World War, requested
from Shapley reprints of the latter's papers on stellar clusters –
but only in 1922, seven years after Shapley began publication.[74]
Indirect support is further suggested by Seeliger himself. In his
last major paper, "On the Investigation of the Stellar System,"
received for publication on 6 March 1920, in which he presented
his most definitive research and conclusions on the structure and
form of the Milky Way system, no reference is made to Shapley's
highly unorthodox and opposing work. And finally, in a letter
written to Kienle in 1922, Shapley makes it clear that Seeliger's
work was relatively unknown among American astronomers, suggesting that Seeliger had few American contacts and perhaps little
regard for the innovative American work:

> I agree with you that we appear to know too little
> concerning the contributions made by Seeliger to the
> problem of galactic structure. There are three insuffi-
> cient reasons for this. Seeliger's publications are not as
> accessible as some. We in America, at least, have been
> influenced and gained greatly by Kapteyn and his school,
> and they have not been quite fair, it now seems to me,
> to Seeliger's work. Seeliger's conclusions [concerning
> his galactic theory], as given in successive editions of
> Newcomb–Engelmanns Popular Astronomy [sic], have
> not remained the same.[75]

[72] See H. Shapley to J. C. Kapteyn, 25 July 1918, and 19 December 1918
(Shapley).
[73] Felix Schmeidler to E. Robert Paul, 18 October 1978 (in author's possession).
[74] H. Shapley to H. Kienle, 11 October 1922 (Shapley).
[75] H. Shapley to H. Kienle, 11 October 1922 (Shapley).

Thus it seems that Seeliger did not engage in either serious investigations or polemics against Shapley's newer approach and his stellar system because he was unaware of the more recent developments being published in the periodicals at the time and because he lacked direct personal contact with those engaged in the debate.

Although it seems to be the case that Seeliger was largely unaware (or at least unfamiliar) with Shapley's views, sometime during 1922 he gained considerable acquaintance with Shapley's ideas, and privately expressed his own views on the matter. Following the IAU meetings in Rome, Shapley began a series of visits with European colleagues during May and June, no doubt to discuss his unorthodox views. In addition to his Leiden conference on May 27 with Kapteyn & company, Shapley also visited Munich. Later correspondence of Seeliger's suggests, however, that it is very doubtful that he actually met with Seeliger, though he most certainly attempted to see Kienle, Seeliger's former student and younger colleague. Unfortunately, Kienle himself was traveling, and the two did not meet at this time. But within a few months, Kienle and Shapley commenced a warm correspondence, which resulted in a mutual advantage to both astronomers. Kienle received directly from Shapley many of the crucial papers Shapley had produced since 1915 dealing with Shapley's "big galaxy," whereas Shapley won an early European convert to his newer system.[76] As a result of Kienle's emerging relationship with Shapley, Seeliger obtained direct access, probably by late 1922, to Shapley's work. Seeliger quickly realized the implications of Shapley's views, and then proceeded, at least informally, to counter Shapley's results. In a highly revealing letter to his lifelong friend and colleague, Max Wolf, who at the time was the director of the observatory at Heidelberg University, Seeliger in part wrote:

> Although some critical observations are in order, not so much in detail but in the final end results, many of the overall results of Shapley appear to me more than doubtful, which also is admitted by some Americans. In my opinion, the parallaxes of the [globular] clusters

[76] H. Kienle to H. Shapley, 21 September 1922, and H. Shapley to H. Kienle, 11 October 1922 (Shapley).

are too small, even if the opinions of Curtis must be considered invalid. Regarding the spiral nebulae, one may come to the same conclusion as I have for more than twenty years regarding the universe, that is very satisfactory to me. Also my solid opinion about the star-counts [receives] favorable comments beyond my fondest expectations.[77]

Thus Seeliger shared Kapteyn's view that the parallaxes to the clusters are perhaps a magnitude too small. Seeliger, along with Curtis and Kapteyn, continued to argue for the island universe theory, which suggested that the spirals were separate galaxies.

More important, however, Seeliger never published his opposition to Shapley's work. He was already seventy-three years of age when Kienle showed Seeliger the research papers Shapley had sent. Moreover, conditions were no longer ideal for serious, sustained research in Germany for a number of years following the war. In a letter to Wolf a few years earlier, Seeliger quietly complained:

> The outlook in our science is rather bleak; one does not know where all these restrictions [imposed by Versailles] will lead to and what will be the result of it all. But German scientists are strong and will not lose.... With our coming generation, however, the future does not look good, as the unprincipled government of Prussia is certain to change everything to our disadvantage. Because of limitations of maintenance and aging, the observatory at Babelsberg [Berlin] is bereft of young talent.[78]

Things had improved little by 1923 when Seeliger again wrote Wolf:

> Most likely you are suffering under the same wretched conditions in Germany as I do. I am afraid that we are facing intolerable conditions in our scientific endeavors in Germany, especially as concerns astronomy, which at this stage demands great support in order to keep us

[77] H. v. Seeliger to M. Wolf, 17 February 1923 (Heidelberg).
[78] H. v. Seeliger to M. Wolf, 10 February 1921 (Heidelberg).

abreast with developments in other countries, especially with America where much is done. . . . The emotional conditions I find myself in I just cannot concentrate much on the real work before me. On top of it the bothersome Directors of the Academy [of Sciences at Munich] get under my skin, and I plan on getting rid of them. . . . I hope you and your loved ones are in the best of health, something I cannot say for myself, since the disgrace and shame [of losing the war] lays too heavily on me. If I was ten years younger I could cope with this national tragedy much better, but the hope of better times is impossible for me. And without hope, life is without meaning and sad.[79]

Thus Seeliger eventually became familiar with Shapley's ideas. Two months before his death in 1924, Seeliger again saw some of Shapley's results, which this time were published in Germany. A Festschrift dedicated to Seeliger appeared with twenty-four papers on contemporary astronomical problems. It was edited by Seeliger's younger colleague Kienle, whom Shapley, of course, considered to be a very capable and competent statistical astronomer (because Kienle agreed with Shapley?) and who had been continuing his mentor's work in statistical cosmology since 1918. This volume included, ironically, a paper by Shapley on the distance determination of the Magellanic Clouds by the use of cepheid variables.[80] More than any other of Seeliger's students, Kienle had worked vigorously in the area of statistical cosmology and eventually rose to become Germany's most outstanding astronomer of the twentieth century. Notwithstanding his training under Seeliger's tutelage, Kienle, as one of Seeliger's last doctoral students (1918), began his career prepared to see Seeliger's views, interpreted as describing the local solar neighborhood, reconciled

[79] H. v. Seeliger to M. Wolf, 17 February 1923 (Heidelberg).

[80] H. Shapley, "The Magellanic Clouds," in *Problem der Astronomie: Festschrift für H. v. Seeliger*, ed. by H. Kienle (Berlin: Springer Verlag, 1924), 438–41. Shapley's distances to the Magellanic Clouds were confirmed independently by others at the time. See R. E. Wilson, "On the Distance of the Large Magellanic Cloud," *Astronomical Journal*, 35 (1924), 183–4, and K. Lundmark, "The Distance of the Large Magellanic Clouds," *Observatory*, 47 (1924), 276–9.

with Shapley's radically different ideas.[81] Writing to Shapley in 1922, Kienle noted:

> It appears to me that Seeliger's results, a limited system from 6,000 to 26,000 lightyears in diameter, should be very sympathetic to you. One needs only to suppose that Seeliger's star-counts refers to the "local cluster".... Especially if your views would prove to be correct, one would have to consider Seeliger's prominently outstanding boundaries of the local cluster; the advantage over Seeliger would exist only in that this boundary is not to be measured with respect to the Milky Way. Rather it would point to a higher order in the systems of the cluster. It would please me greatly, if you, dear sir, would express your opinion in this matter.[82]

Shapley answered Kienle, and acknowledged that Seeliger was, in Shapley's opinion, too little known among American astronomers.[83] Thus the apparent contradictory nature of the Festschrift was in Kienle's judgment an attempt at rapprochement between ostensibly incompatible stellar models.

Compared to Seeliger's reaction, the response in Sweden to Shapley's newer developments was more in line with Kienle's response (and also partially with van Rhijn's response). Charlier was a relative newcomer to the field compared to both Kapteyn and Seeliger, and his school not only had less to lose professionally in accepting the newer views, but in time it also actively engaged in research in support of Shapley. Two years before Shapley announced his radical cosmology in 1918, Charlier hinted at the possibility of an enlarged stellar system in an important study using the statistical methods of stellar astronomy. Examining the distances and distributions of the B-spectral type stars (those of high absolute magnitude), Charlier found that their center coincided with the center of our stellar universe at a distance of

---

[81] Kienle was not alone among Seeliger's students to share his optimism for Shapley's newer views; see for instance, P. ten Bruggencate, "Note on the Structure of Shapley's Larger Galactic System," *Bull Astron Neth*, 4 (1928), 198–201, and P. ten Bruggencate, *Sternhaufen: Ihr Bau, ihre Stellung zum Sternsystem und ihre Bedeutung für die Kosmologie* (Berlin, 1927), part I.

[82] H. Kienle to H. Shapley, 21 September 1922 (Shapley).

[83] H. Shapley to H. Kienle, 11 October 1922 (Shapley).

THE SECOND ASTRONOMICAL REVOLUTION

about 90 parsecs (300 light years) from the Sun.[84] This in itself was a significant departure from the current traditional views.

Within a year, it was clear that Charlier was taking Shapley's work on stellar clusters seriously. After deriving a sidereal system about 9,000 parsecs in diameter, Charlier concluded with this qualification:

> The value obtained for the scale is, however, subject to considerable uncertainty, all the more as other consid-erations [i.e., Shapley's] lead to much higher (10 times as high) values of the scale. It is to be expected that the variable stars of short period [i.e., cepheids], so abundant in the globulars, will give material for a surer deter-mination of the scale.[85]

In an unpublished lecture on the structure of the Milky Way system, delivered before the Swedish Astronomical Association in May 1920, Charlier suggested that the essential features of Shapley's cosmology might in fact be closer to the actual state of affairs.[86] And during Shapley's 1922 European tour, the Swedes invited Shapley to Stockholm (expenses paid) to lecture on galac-tic dimensions.[87]

THE BEGINNING OF A NEW CONSENSUS

A perusal of Shapley's personal correspondence makes it clear that Shapley was so confident of his "vaster universe," as Kapteyn facetiously called it, that he personally lobbied for his case when-ever the opportunity arose.[88] One of the first opportunities to do so was presented to Shapley by Charles G. Abbot, secretary of the National Academy of Sciences, when the latter suggested a confrontation between Shapley and Heber D. Curtis early in 1920

---

[84] C. V. L. Charlier, "Studies in Stellar Statistics – The Distances and the Distribution of the Stars of the Spectral Type B," *Meddelanden Fran Lunds Astronomiska Observatorium* (Serie II), no. 14 (1916), 31, 104 (108 pp.).

[85] C. V. L. Charlier, "Stellar Clusters and Related Celestial Phenomena – Stud-ies in Stellar Statistics," *Lunds Medd.* (II), no. 19 (1918), 41 (54 pp.).

[86] Charlier, "Stjarnrakningen pa Lunds observatorium och Vintergatans byggnad" (11 May 1920); also see Shapley and Curtis, "The Scale of the Universe," p. 174.

[87] H. Shapley to G. R. Agassiz, 19 June 1922 (Shapley).

[88] J. C. Kapteyn to G. E. Hale, 3 March 1920 (Hale).

over the scale of the Universe.[89] Though Shapley was eager to accept the challenge to debate Curtis (or Campbell, or whomever), Shapley was concerned about the prospect of facing an experienced public speaker, not because of the technical issue on the agenda but because an appointment as the director of the Observatory at Harvard was pending and Shapley was the frontrunner. Although his performance against Curtis could hardly have been more disastrous, eventually Shapley went to Harvard and in October of 1921 he was offered the directorship.[90]

His advocacy of his big universe, however, remained unabated. For example, in a September 1920 letter to Hale, who had already accepted Shapley's view on the matter, Shapley commented on Walter S. Adams's evidence for and subsequent acceptance of giant stars in globular clusters, thereby reinforcing Shapley's claim for large distances to clusters. "I am greatly pleased to hear (and so are others who believe in the big Galaxy), that [Adams] now believes spectroscopic evidence proves these stars to be giants. It will have great weight in many obstinate [?] corners. It should be, I think, the last straw that breaks the Campbell's back."[91] The latter reference, of course, was to the Lick Observatory, where Campbell, Curtis, and their staff remained supporters of the smaller sidereal universe. Thus while Shapley was being seriously considered for the Harvard directorship, Hale could write Harvard University President A. Lawrence Lowell with confidence that "Dr. Shapley's ... views on the structure and extent of the universe have received wide acceptance, though they are still opposed in some quarters."[92] Shapley did not relent, and whenever others publicly questioned his results in the technical literature or suspended their judgment of his conclusions, Shapley would not hesitate to state his case to them directly and suggest, as he did to the Dutch-American astronomer Willem J. Luyten in 1922, that "rather than publish comments that might [otherwise] unjustly throw a shadow on ... your reputation" it would be well

[89] Charles G. Abbot to G. E. Hale, 18 February 1920; H. Shapley to G. E. Hale, 19 February 1920 (Hale).
[90] O. Gingerich, "How Shapley came to Harvard or, Snatching the Prize from the Jaws of Debate," *Journal for the History of Astronomy*, 19 (1988), 204–6 (201–7).
[91] H. Shapley to G. E. Hale, 12 September 1920 (Hale).
[92] G. E. Hale to A. Lawrence Lowell, 11 December 1920 (Hale).

for you to reconsider your "suspended judgment."[93] Although resistance came from the Lick astronomers, the main opposition, as we have argued, came from Kapteyn and the Dutch statistical cosmologists.

[93] H. Shapley to Willem J. Luyten, 18 July 1922 (Shapley).

9

CONCLUSION: RESEARCH PROGRAMS IN TRANSITION

The research program initiated a century-and-a-half earlier by William Herschel that determined the arrangement of the sidereal universe by counting stars (and, later, measuring stellar magnitudes) had failed in a fundamental way. Even the sophisticated mathematico-statistical techniques developed by Kapteyn, Seeliger, and others were unable to resolve the features of a fundamentally more complex universe. In the last analysis, however, the methodologies and models of classical statistical astronomy were seriously inadequate for the task for which they were created. As a scientific research program of global dimensions, statistical cosmology had reached its zenith with the emergence of Shapley's program of analyzing globular clusters. After Kapteyn's death in July 1922 and Seeliger's in December 1924, virtually nothing was written of a technical nature in defense of the classical models. Still, even with the near explicit approval of Charlier and the apparent acquiescence of Seeliger, the conflict over dimensions and distances had not been settled by the mid-1920s.

THE "NEW ASTRONOMY"

Much had been written between 1918 and 1922 about the conflict over the classical methods and sidereal models of statistical astronomy and the new cosmology. On New Year's Day 1925, however, Edwin P. Hubble (1889–1953), an American astronomer and colleague of Hale's at Mount Wilson, had a paper delivered at the annual meeting of the American Astronomical Society that overwhelmingly supported the enormous distances Shapley had suggested. Following Shapley in using cepheid variables as galactic measuring rods, Hubble obtained distances to the spiral nebulae M31 and M33 (Andromeda) of about 300,000 parsecs

(950,000 light years).[1] Although his results immediately confirmed the island universe theory advocated by Shapley's rival Curtis, they also greatly strengthened the use of cepheids as indicators of distances and thus helped confirm the great size of the Milky Way system advocated by Shapley. So, after 1925 it became increasingly difficult to think in terms of a restricted universe in the classical sense of statistical cosmology. As a result, cepheid variables soon became universally accepted as the astronomer's measuring rod.

It was now possible, as Kienle had done, to posit a smaller system such as Kapteyn's or Seeliger's as defining the *local* system of stars and to consider spiral nebulae as systems comparable to the Milky Way at distances in excess of 300,000 parsecs (or about a million light years).[2] As a result, not only did the Milky Way Galaxy shrink as the Universe grew with Hubble's discovery, but other features in the research program defined by statistical astronomy were also profoundly changed during the 1920s.

A major theoretical support for the larger stellar system, in

---

[1] See Edwin P. Hubble, "Cepheids in Spiral Nebulae," *Observatory*, 48 (1925), 139–42; and E. P. Hubble, "Distances of the Andromeda Nebula," *Popular Astronomy*, 33 (1925), 143. In contrast, van Maanen had earlier derived a distance to Andromeda of only 290 parsecs (900 light years); see A. van Maanen, "The Parallax of the Andromeda Nebula," *Astronomical Society of the Pacific, Publications*, 30 (1918), 307. For his paper, Hubble shared the joint award of the $1,000 prize given for the most outstanding paper at the annual meeting of the American Astronomical Society. His conclusions, in the words of Shapley's mentor Henry Norris Russell, were accepted, "thus bringing confirmation to the so-called island universe theory." See "33rd Meeting of the AAS," *Popular Astronomy*, 33 (1925), 159. It is interesting – and somewhat ironic – that Hubble used Shapley's modification of the period-luminosity constants for cepheid variables to establish the island universe dimensions of the spirals; see W. W. Campbell, "Do We Live in a Spiral Nebulae?" *Popular Astronomy*, 34 (1926), 174–81. For a discussion of these details, see Richard Berendzen and Michael Hoskin, "Hubble's Announcement of Cepheids in Spiral Nebulae," *Astronomical Society of the Pacific, Leaflet*, no. 504 (1971), and Richard Berendzen, Richard Hart, and Daniel Seeley, *Man Discovers the Galaxies* (New York: Science History Publ., 1976); there are also some relevant details in Robert W. Smith, *The Expanding Universe: Astronomy's "Great Debate," 1900–1931* (Cambridge: Cambridge University Press, 1982).

[2] For examples of such smaller galactic systems within the larger Hubble Universe, see C. Wirtz, "Die Spiralnebel und die Struktur des Raumes," *Scientia*, 38 (1925), 303–14; and H. D. Curtis, "The Unity of the Universe," *Royal Astronomical Society of Canada*, 22 (1928,) 399–412.

which the Sun was radically displaced from the center, came from entirely different considerations. Ever since Kapteyn's announcement of star-streaming in 1904, astronomers had largely failed to explain the physical basis of this phenomenon. Both Eddington's and Schwarzschild's models provided a theoretical analysis of star-streaming, but neither succeeded in explaining the underlying cause of these preferential stellar drifts. Although Kapteyn changed his position later, he suggested initially that star-streaming was the result of two streams of stars moving through one another in precisely opposite directions. In 1924 Charlier mathematically demonstrated that the distribution of the velocities of the stars, though in complete accordance with the general theory of statistical mechanics suggested by Eddington and Schwarzschild, was absolutely "in contradiction to the existence of the proposed two star streams of Kapteyn."[3] Meanwhile Kapteyn and James Jeans attempted to explain star-streaming along dynamic lines within the smaller Kapteyn Universe.[4] As we have seen, Kapteyn suggested a somewhat eccentrically located Sun, whereas Jeans, who still favored a Sun-centered universe,[5] interpreted star-streaming as being due to the circular motions in opposite directions rotating around the center of the stellar system.

These latter attempts failed, however, to take into account certain anomalous observations, particularly the asymmetrical drift of high velocity stars discovered by G. B. Strömberg in 1923.[6] Strömberg suggested that in order to account for high velocities through dynamic and gravitation forces of the system as a whole,

[3] C. V. L. Charlier, "Do the Star Streams of Kapteyn Exist?" *Astronomical Society of the Pacific, Publications*, 36 (1924), 213 (212–15).

[4] J. C. Kapteyn, "First Attempt at a Theory of the Arrangement and Motion of the Sidereal System," *Astrophysical Journal*, 55 (1922), 316–18 (302–28); and J. Jeans, "The Motions of the Stars in a Kapteyn-Universe," *Royal Astronomical Society, Monthly Notices*, 82 (1922), 122–32.

[5] J. Jeans, "The Origin of the Solar System," in *Probleme der Astronomie: Festschrift für H. v. Seeliger*, ed. by H. Kienle (Berlin: Springer, 1924), p. 1 (1–24).

[6] The phenomenon of high velocity stars, unknown to Kapteyn and Jeans when they published their papers in 1922, first became known in 1924; see G. B. Strömberg, "The Asymmetry in Stellar Motions and the Existence of a Velocity-Restriction in Space," *Astrophysical Journal*, 59 (1924), 228–51, and G. B. Strömberg, "The Asymmetry in Stellar Motions as Determined from Radial Velocities," *Astrophysical Journal*, 61 (1925), 363–88.

a much larger system than the one posited by Kapteyn must be assumed.[7] All high velocity stars tended to rotate in the same direction perpendicular to the galactic center of Shapley's model.

Beginning in 1925 the Swedish astronomer Bertil Lindblad (1895–1965) developed a theory that explained these various phenomena in terms of Shapley's larger galactic system. In this theory the complete galactic system was divided into a number of concentric, ellipsoidal subsystems, each rotating with its own velocity about a common axis perpendicular to the galactic plane. The innermost system contained the Milky Way, including our local system, and rotated such that the highest velocity was the most flattened. Each successive subsystem rotated with less speed and was therefore less eccentric. The outer subsystem, defining the globular clusters, rotated the slowest and hence was nearly spherical in shape. The entire galactic system was coextensive with Shapley's system.[8]

Given this galactic structure, Lindblad was able to explain the asymmetry of the high velocity stars discovered by Strömberg. According to Lindblad's theory, as a group these stars must move in a direction opposite to galactic rotation and generally perpendicular to the galactic center exactly as Strömberg had observed. As a result of his theory, Lindblad placed the center of the larger system at galactic longitude 330°. The center of the system of globular clusters found by Shapley was in the direction of Sagittarius at longitude 325°.[9] As a result, Lindblad was able to show that star-streaming was an indirect consequence of galactic rotation.[10]

Lindblad went on to demonstrate that the total mass calculated from the Kapteyn Universe produced a gravitational potential too weak to retain the globular clusters and certain of the high velocity

[7] Strömberg, "The Asymmetry in Stellar Motions and the Existence of a Velocity-Restriction in Space," p. 249.

[8] B. Lindblad, "Star-Streaming and the Structure of the Stellar System," *Meddelanden Fran Upsala Astronomiska Observatorium* (hereafter *Upsala Medd.*), no. 3 (1925), 1–2 (8 pp.).

[9] Ibid., p. 2; and B. Lindblad, "On the State of Motion in the Galactic System," *Royal Astronomical Society, Monthly Notices*, 87 (1927), 553 (553–64).

[10] B. Lindblad, "On the Cause of Star-Streaming," *Astrophysical Journal*, 62 (1925), 191–7.

variable stars. Because both these stellar objects were found in large numbers, either they were evolved at a rate sufficient to compensate for those that escaped or they were permanent members of the larger galactic system bound by gravitational forces much larger than allowed by the Kapteyn Universe. The enormous size of the globular clusters mitigated against the first alternative. Hence Lindblad favored the latter view that implied a much more massive galactic system than Kapteyn, Seeliger, or any of the statistical cosmologists proposed.[11] From dynamic and gravitational considerations, therefore, Lindblad's theory indirectly confirmed Shapley's galactic model and thus helped to undermine the classical approach to cosmology.

In 1927 Jan Oort (1900–1992), who became Holland's most influential and gifted astronomer, produced observational evidence that helped confirm this view. Oort sought to verify the fundamental hypothesis of Lindblad's theory, namely, that of the rotation of the galactic system around a point near the center of the system of globular clusters.[12] In a rotating system there are basically two explanations for dynamic equilibrium. On the one hand, if the mass is concentrated at a point, the gravitation force would be proportional to the inverse square of the distance, causing the innermost stars to rotate at a greater velocity than those further out. On the other hand, if the mass is uniformly distributed, the stars would rotate with the same angular velocity because the force would vary directly as the distance between stellar objects. Not only did the theoretical results of the statistical

---

[11] See B. Lindblad, "On the Dynamics of the System of Globular Clusters," *Upsala Medd.*, no. 4 (1926), 8 pp.; idem, "Star Streaming and the Structure of the Stellar System," *Upsala Medd.*, no. 6 (1926), 6 pp.; and idem, "Cosmogonic Consequences of a Theory of the Stellar System," *Upsala Medd.*, no. 13 (1926), 15 pp.

[12] Jan H. Oort, "Observational Evidence Confirming Lindblad's Hypothesis of a Rotation of the Galactic System," *Bulletin of the Astronomical Society of the Netherlands*, 3(120) (1927), 275–82. Also see J. Oort, "Investigation Concerning the Rotational Motion of the Galactic System, together with new Determination of Secular Parallaxes, Precession and Motion of the Equinox," *Bulletin of the Astronomical Society of the Netherlands*, 4 (132) (1927), 79–89, in which Oort sought to give definitive discussion of rotations shown by proper motions. For a first-hand account of Oort's derivation of his famous mathematical theory, see "Bart Bok Interview" (American Institute of Physics), 15 May 1978, pp. 10–11.

cosmologists mitigate against the second alternative, but the large negative velocities, relative to the Sun, of most of the high velocity stars, also favored the first alternative. Consequently, Oort theoretically developed the first possibility, thus helping to confirm Lindblad's theory.

Known as "differential galactic rotation," Oort's analysis explained the large negative velocities of the high velocity stars. Simply put, these stars, which are further from the galactic center than the Sun, move more slowly with respect to the center. (Today, it is recognized that this phenomenon is due more to the stars' high eccentricity than to their great distance from the Sun.) Because the Sun is closer to the center and therefore has a higher angular velocity, it appears, as seen from the Sun, that the high velocity stars are receding at a "high velocity." Furthermore, Oort demonstrated that when stars have motions deviating from exact circles, their velocities relative to the Sun will show a preferential direction toward the galactic center. This was precisely the effect generated by the phenomenon of star-streaming. Oort concluded with a discussion of the distance to the galactic center. He had obtained his results as the effect of rotation around a central mass located at $325°$ longitude. The distance from the Sun to this massive center (about 10^{10} solar masses) was on the order of 6,600 parsecs, a distance about one-third of that posited by Shapley. Again these results, Oort showed, verified the general features of Shapley's galactic system. On the strength of these developments, the Lindblad–Oort hypothesis, the name that has come to identify differential galactic rotation, began almost immediately to receive support throughout the astronomical community.[13]

There remained the difficulty, however, which Oort pointed out, of why the large mass in the direction of Sagittarius was

[13] See, for instance, J. S. Plaskett, "The Rotation of the Galaxy," *Royal Astronomical Society, Monthly Notices*, 88 (1928), 395–403; J. S. Plaskett and J. A. Pearce, "The Rotation of the Galaxy," *Astronomical Society of the Pacific, Publications*, 41 (1929), 251; and A. S. Eddington, *The Rotation of the Galaxy* (Oxford: Oxford University Press, 1930), 30 pp. In 1930 Shapley himself began a series of studies of the galactic center; see H. Shapley and H. W. Swope, "Studies of the Galactic Center. II. Preliminary Indication of a Massive Galactic Nucleus," *National Academy, Proceedings*, 14 (1928), 830–4.

unobservable. Kapteyn and van Rhijn had derived almost negligible density near 6,600 parsecs in their 1920 model. In Oort's opinion, the answer was already implied in Olbers's suggestion of interstellar absorption. "The most probable explanation," wrote Oort, "is that the decrease of density in the galactic plane indicated for larger distances is mainly due to obscuration by dark matter."[14]

As we have abundantly seen, stellar astronomers were extremely sensitive to the possibility of an absorbing medium.[15] Historically the attention of astronomers was first directed to this problem by statistical investigations of the spatial distribution of the stars. In most instances, they expressed extreme caution as a final caveat about the potentially disastrous consequences that the existence of interstellar light absorption would pose for their stellar models. In the case of the Lindblad–Oort theory, however, the existence of such a medium was actually demanded by their theory.

Even though conclusive observational evidence for the existence of light absorption did not appear until 1930, most stellar astronomers as early as 1910 were convinced – though uneasily – of its existence. Until the 1920s most of these studies were made using methods developed in stellar statistics.[16] The results of these

[14]   Oort, "Observational Evidence Confirming Lindblad's Hypothesis of a Rotation of the Galactic System," p. 281.
[15]   For an informative historical review of the evidence for interstellar light absorption, see F. H. Seares, "The Dust of Space," *Astronomical Society of the Pacific, Publications*, 52 (1940), 80–115. A more recent examination of this problem has alluded to the role that statistical astronomy has played in the general problem of interstellar absorption; see D. Seeley and R. Berendzen, "The Development of Research in Interstellar Absorption, c. 1900–1930," *Journal for the History of Astronomy*, 3 (1972), 52–64 and 75–86, particularly pp. 75–80.
[16]   For the most important early attempts using statistical techniques, see G. C. Comstock, "Stellar Luminosity and the Absorption of Star Light," *Astronomical Journal*, 24 (1904), 139–43; J. C. Kapteyn, "Remarks on the Determination of the Number and Mean Parallax of Stars of Different Magnitudes and the Absorption of Light in Space," *Astronomical Journal*, 24 (1904), 115–23; H. Seeliger, "Ueber die räumliche Verteilung der Sterne im schematischen Sternsystem," *München Ak. Sber.*, 41 (1911), 413–61; J. Halm, "On the Question of Extinction of Light in Space and the Relations between Stellar Magnitudes, Distances and Proper Motions," *Royal Astronomical Society, Monthly Notices*, 77 (1917), 243–80; J. Halm, "Statistical Investigations of the Distribution of the Stars and their Luminosity," *Royal Astronomical Society, Monthly Notices*, 80 (1919), 162–98; A. Kopff, "Ueber die Absorption im Weltenraum," *Zeitschrift für Physik*, 17 (1923),

methods varied widely, and in most cases required the acceptance of assumptions that were themselves highly problematic.

During the late 1920s, with the rise in significance of globular clusters for an understanding of the entire galactic system, some astronomers attempted to examine this basic issue using clusters rather than individual stars. Unlike most single stellar objects, the distances to the clusters could be determined by several independent methods. Moreover, the distances were enormous, varying from 6,000 to 60,000 parsecs. At these distances the effects of any absorption should be easily detectable. The basic assumption of the cluster approach was that the globular clusters were mostly of comparable sizes. After achieving several conflicting results using clusters to determine the "absorption coefficient,"[17] which gives the light extinction in magnitudes per kilo-parsec, the Swiss-American astronomer Robert J. Trumpler (1886–1956) published in 1930 what has since become recognized as the definitive argument for the existence of interstellar light absorption.[18]

Trumpler's method is really quite simple and straightforward: determine the distances to the globular clusters by two independent methods, one of which assumes no spatial absorption. Consequently, he based his methods on stellar magnitudes and the

279–86; H. Kienle, "Die Absorption des Lichtes im interstellaren Raume," *Jahrbuch der Radioaktivitat und Elektronik*, 20 (1922), 1–12; H. Kienle, "Die Absorption des Lichtes und des Grenze des Sternsystem," *Zeitschrift für Physik*, 20 (1924), 388–93; P. ten Bruggencate, "Die Bestimmung einer allgemein Absorption des Lichtes im Weltraum," *Zeitschrift für Physik*, 57 (1929), 631–7; and C. Shalen, "Zur Frage einer allgemeiner Absorption des Lichtes im Weltraum," *Astronomisches Nachrichten*, 236 (1929), 249–58.

[17] See K. Lundmark, "The Motions and the Distances of Spiral Nebulae," *Royal Astronomical Society, Monthly Notices*, 85 (1925), 865–94; P. van Rhijn, "On the Absorption of Light in Space Derived from the Diameter-parallax Curve of Globular Clusters," *Bulletin of the Astronomical Institutes of the Netherlands* 4 (1928), 123–8; and H. Shapley and M. Ames, "The Coma-Virgo Galaxies, I: On the Transparency of Intergalactic Space," *Harvard Astronomical Observatory Bulletin*, no. 864 (1929), 1–6.

[18] In 1930 Trumpler published three papers on this subject that have since become classics: see R. J. Trumpler, "Preliminary Results on the Distances, Dimensions and Space Distributions of Open Star Clusters," *Lick Observatory Bulletin*, 14 (1930), 154–88; idem, "Absorption of Light in the Galactic System," *Astronomical Society of the Pacific, Publications*, 42 (1930), 214–27; and idem, "Spectrophotometric Measures of Interstellar Light Absorption," *Astronomical Society of the Pacific, Publications*, 42 (1930), 267–74.

sizes of clusters, because interstellar absorption would diminish the brightness of the stars but not the dimensions of the clusters appreciably. Trumpler first examined the apparent magnitudes and spectral types of the members of 100 clusters. Using the Hertzsprung–Russell diagram and the magnitude–distance relationship, he then derived mean distances to these clusters. This method, which requires the assumption of no light absorption whatsoever, had become a standard technique for the distance measurement of individual stars after the formulation of the Hertzsprung–Russell diagram. His second method, which would account for the existence of any absorption, was based on the general structure of the globular clusters. Here too he began with the fundamental assumption used in clusters studies, namely, that their diameters are roughly equivalent. Recognizing the many differences in the structure of globular clusters, however, Trumpler made an important qualification of this basic supposition:

> In view of the great diversity in cluster formations it seemed a priori likely that the real space dimensions are not the same for all of them but that they depend on the constitution of each cluster. The open clusters were therefore classified according to the degree of star concentration toward the center and according to the number of stars contained in them. As expected, the linear diameter ... was found to be correlated with both of these characteristics. Making the assumption that clusters of similar constitution have on an average everywhere the same dimensions, it is possible to determine the distance of a cluster ... by comparing its angular diameter with the mean linear diameter of the subclass to which it belongs.[19]

These "diameter distances" were thus based on the angular diameters of the clusters; double the distance to a cluster and it appears half as large. Consequently, this empirical evidence would not reflect the existence of an absorbing medium. In other words, this method gave the real distances to the clusters. If there was no absorption, the two distance measures would presumably be identical. His results indicated, however, that the direct distance

[19] Trumpler, "Absorption of Light in the Galactic System," p. 217.

scale was systematically smaller. Assuming an absorption uniform within the region defined by the clusters, Trumpler derived an absorption value of 0.7 magnitudes per 1,000 parsecs.

Trumpler's results were quickly confirmed by others and have since become well established.[20] Trumpler's significant work confirmed Oort's suggestion that the large central mass in the direction of Sagittarius was obscured by an absorbing material. Of even greater moment, though, the widespread effects of interstellar light absorption became the "achilles heel" of the classical statistical models. Ultimately the smaller universe of Kapteyn and Seeliger was predicated on the counts and brightnesses (apparent magnitudes) of the stars. Beyond the outer limits of the classical systems, about 9,000 parsecs, stellar objects recede into the milky clouds of the galactic system. Objects in the galactic center (at about this distance of 9,000 parsecs from the Solar System) were therefore essentially invisible.

It would be incorrect to suggest that stellar statistics was rejected in toto; rather it was deemed wholly inadequate as the approach for determining the architecture of the much larger Shapley system. Henceforth, astronomers began talking of the "local system," for which statistical astronomy had proved itself of great worth.[21] Even Shapley recognized as much: "In the past, Seeliger, E. C. Pickering, Kapteyn, and many others have made extensive investigations of the distribution of stars, for the problem is recognized as fundamental in the study of galactic structure."[22]

[20] P. van de Kamp. "On the Absorption of Light in Space," *Astronomical Journal*, 40 (1930), 145–59; K. F. Bottlinger and H. Schneller, "Ueber die interstellar Absorption immer-halb der Milchstrasse," *Zeitschrift für Physik*, 1 (1930), 339–42; and Seares, "The Dust of Space," pp. 100–1.

[21] In 1928 the astronomer Hector Macpherson, Jr., wrote: "Shapley's work has revolutionized our idea of the scale of space. His universe – 300,000 light years [90,000 parsecs] in diameter – far exceeded the universe as conceived by the astronomers of the later 19th and earlier 20th centuries, who, as Shapley showed, had really been concentrating upon a local cluster, under the natural impression that it was co-extensive with the Universe." See H. Macpherson, "The Frontiers of the Universe," *Observatory*, 51 (1928), 321 (315–22).

[22] H. Shapley and A. J. Cannon, "Summary of a Study of Stellar Distribution," *American Academy, Proceedings*, 59 (1924), 215 (215–31); also see H. Shapley, "The Distribution of the Stars," *Popular Astronomy*, 32 (1924), 418–23, in which Shapley presents some original research in classical stellar distributions.

Through the 1920s, van Rhijn, Seares, Pannekoek, Charlier, and others continued to use stellar statistics in their investigations into various problems dealing with the restricted local system.[23]

By 1930, however, Shapley's radical cosmology, combined with the Lindblad–Oort theory of differential galactic rotation and Trumpler's verification of the existence of an interstellar absorbing medium, included the theoretical and empirical work that finally caused the demise of the classical theories and, combined with Hubble's highly innovative work in extragalactic astronomy, ushered in a "golden era" in astronomy. Once again, as with William Herschel's theories much earlier, the statistical cosmologies were found to be inadequate for the task at hand. This time, however, even the general approach to an understanding of the sidereal problem was, in the final analysis, also found to be inadequate.

CONCLUDING COMMENTS

Whereas scientists who are actually engaged in doing science are interested in solving problems and in constructing theories of explanation, historians and philosophers of science, who focus on the intellectual dimensions of science, are interested, among

[23] References to this problem in the technical literature are numerous. For some of the more significant papers, see A. Pannekoek, "The Local Starsystem," *Koninklijke Akademie van Wetenschappen te Amsterdam. Proceedings of the Section of Sciences* (hereafter *Amst. Ak. Proc.*), 24 (1) (1922), 56–63; idem, "Researches on the Structure of the Universe," *Astronomical Institute of the University of Amsterdam, Publications,* 1 (1924), 121 pp.; K. Malmquist, "Researches on the Distribution of the Absolute Magnitudes of the Stars," *Lund Medd. II,* no. 32 (1924), 77 pp.; F. H. Seares, "Remarks on the Luminosity and Density Functions," *Astrophysical Journal,* 59 (1924), 11–29; F. H. Seares, "The Form of the Luminosity Function," *Astrophysical Journal,* 59 (1924), 310–38; P. v. Rhijn, "On the Frequency of the Absolute Magnitudes of the Stars," *Groningen Publications,* no. 38 (1925), 77 pp.; K. Malmquist, "A Contribution to the Problem of Determining the Distribution in Space of the Stars," *Lund Medd.* (I), no. 106 (1925), 12 pp.; C. V. L. Charlier, *The Motion and Distribution of the Stars* (Berkeley: University of California Press, 1926); A. Pannekoek, "Researches on the Distribution of the Stars in Space," *Amst. Ak. Proc.,* 32 (1) (1929), 13 pp.; and A. Pannekoek, "Research on the Structure of the Universe," *Astronomical Institute of the University of Amsterdam, Publications,* no. 2 (1929), 87 pp.

other issues, in the role that problem solving plays in the scientific enterprise and in how theories are constructed. Whether they are scientists, historians, or philosophers, virtually all agree that in the process of solving problems and building theories science changes. In fact, no one disputes the claim that science is in a continuous state of flux. Among historians and philosophers of science, however, there is no established consensus for dealing with the mechanism of scientific change and the growth of science. Some emphasize continuities, that science is a continuous process of accretion and even progress, whereas others stress revolutions and discontinuities, that subsequent scientific theories or paradigms are mutually incompatible. In the present book both continuities and discontinuities are clearly at work, with the difference depending upon one's position – both national and generational – within the astronomical community. Neither extreme position, however, seems to dominate. Thus there appears to be a continua of scientific possibilities.

The issue that separates ideas about whether science changes is whether we can provide a rational conception for the process of that change. Although no one seriously disputes the observation that much more is known now than, say, in the time of Copernicus 450 years ago, no one has been able to construct a conceptual scheme that provides for a *rational* explanation for the growth of science. To be sure, science changes; but successive scientific programs do not appear to be commensurate. Thus, is it the case that there is no rational way to account for that change? In one way or another, perhaps science is fundamentally an irrational enterprise.[24]

The fact remains, however, that science changes – and grows! How might one account for that change? If no one has been capable of providing a rational scheme to account for that change, perhaps it is the case that the emphasis on the discontinuities of scientific change is the wrong area to explore. I agree that questions of scientific continuity and discontinuity are significant and interesting for understanding the scientific enterprise. I also believe,

[24] David C. Stove, *Popper and After: Four Modern Irrationalists* (New York: Pergamon Press, 1982).

however, that those concerns are misplaced. In order to understand the growth of science and to provide a rational explanation for that growth, we need to look elsewhere.[25]

Perhaps of greater interest than the details of this branch of astronomy, however, is the significance of this study for the historical development of science. Specifically, what do scientists actually do, and how can one account for the growth of science? Although there were several clearly different approaches to the sidereal problem, the change in consensus from the older statistical cosmology of Kapteyn and Seeliger to this newer approach reflected different responses.

By focusing on developments in stellar astronomy, galactic theory, and pre-relativistic cosmology during the early decades of the present century, we have emphasized the importance of what the astronomers are actually doing qua scientists: to wit, solving problems! As with nearly all scientific disciplines, the raison d'être of astronomy has been and continues to be in its ability to solve problems effectively. Although our purpose has not been to develop

---

[25] During the past two decades, a number of historians and philosophers of science have offered a compelling case for change in science by arguing that only *historicist* theories of scientific change – those grounded deeply in the historical record – are capable of providing an understanding of scientific progress. In addition to Thomas S. Kuhn, *The Structure of Scientific Revolutions* (Chicago: University of Chicago Press, 1970, 2nd ed.), see Larry Laudan, *Progress and Its Problems: Toward a Theory of Scientific Growth* (Berkeley: University of California Press, 1977), Larry Laudan, *Science and Values* (Berkeley: University of California Press, 1984), Arthur Donovan, Larry Laudan, and Rachael Laudan, eds., *Scrutinizing Science: Empirical Studies of Scientific Change* (Dordrecht; D. Reidel Publ., 1988), Stephen Toulmin, *Human Understanding* (Oxford: Clarendon Press, 1972), Gerald Holton, *Thematic Origins of Scientific Thought* (Cambridge, Mass.: Harvard University Press, 1973), Wolfgang Stegmüller, *The Structure and Dynamics of Theories* (Berlin: Springer-Verlag, 1976), Gerald Holton, *The Scientific Imagination: Case Studies* (New York: Cambridge University Press, 1978), Imre Lakatos, *Philosophical Papers*, Vol. I. *The Methodology of Scientific Research Programmes* (New York: Cambridge University Press, 1978), Paul Feyerabend, *Rationalism, Realism and Scientific Method: Philosophical Papers*, Vol. I (New York: Cambridge University Press, 1981), I. Bernard Cohen, *Revolution in Science* (Cambridge, Mass.: Harvard University Press, 1985), Ronald N. Giere, *Explaining Science: A Cognitive Approach* (Chicago: University of Chicago Press, 1988), and David L. Hull, *Science as a Process: An Evolutionary Account of the Social and Conceptual Development of Science* (Chicago: University of Chicago Press, 1988).

a theory of scientific change that accounts for the growth of science, we have constantly emphasized the importance of recognizing the different kinds of problems astronomers examine. The problems that are particularly significant are those that define large areas of research and lead to theories and conceptual schemes that have a large explanatory expanse. Consequently, we focused on research problems that eventually led to increasing success in research traditions. During the years covered in this study, astronomy has dealt with three different kinds of problems – empirical, conceptual, and methodological.

Although our primary purpose has been to reconstruct the development of statistical cosmology during the period roughly from 1890 to 1924, we have focused on what that history tells us about the nature of science as well as how these events effected the emergence of the astronomical revolution of the 1920s. In so doing, we have equally explored the nature, interrelatedness, and status of the scientific problems, even as they changed, that came to define this research tradition.

The architecture of the stellar system remained the central problem of statistical astronomers during the entire period covered by this book. The various problems entailed in this scientific discipline, however, changed considerably. In 1880, stellar astronomers were still interested in traditional empirical problems whose conceptual framework had been detailed a century earlier by William Herschel. By the early 1920s, the very conceptualization of the emerging cosmology had begun to shift radically from these statistical cosmologies and had become increasingly refined by a set of hitherto unsolved and newly articulated problems. As a result, the astronomical scale of organization had expanded enormously from a cosmology limited by the direct measure of data to cosmologies whose data reflected a galactic and even a multigalactic perspective.

During the period of this study, both stellar statistics and statistical cosmology were characterized by the accumulation of massive amounts of empirical data. At various times and on various issues, the data were at once both dictated and driven by mathematico-statistical theories. Although considerable progress was made, indeed the fundamental equation of stellar statistics remains today a key concept in the subdiscipline known as galactic astronomy, statistical cosmology was totally eclipsed by the new astronomy of the 1920s.[26] The change from one research tradition (statistical cosmology) to another (the new astronomy), however, was not fundamentally brought about by the persistence of any empirical, conceptual, or theoretical anomalies. To be sure, the 1920s witnessed major revolutionary changes in astronomy, but these changes emerged in response to other issues and not to a fundamental failure in statistical cosmology. Thus, even though the possibility of interstellar absorption remained a serious problem throughout much of the period covered by this book, in itself absorption was not an anomaly. Although it eventually proved to be the achilles heal of the classical statistical cosmologies, only in concert with the investigations of Shapley, Lindblad, Oort, and Hubble, was the older model eventually discarded.

The astronomers who were centrally featured in this study engaged problem solving with a vengeance, not to discredit theories but rather to construct explanations that expanded the scope of understanding. To be sure, stellar statistics and statistical cosmology grew increasingly more effective in problem solving. But these statistical approaches to problems of cosmology did not engage nature *directly*! The statistical theories were primarily models *organizing* empirical data; in and of themselves, they did not fundamentally reflect underlying *physical* relationships found within nature. As a result, statistical cosmology never developed models that mapped well onto nature. As a science, statistical cosmology developed a highly successful program of problem solving: Models reflecting the astronomical observations were quite successful and a significant research tradition emerged that for a generation (or two) became dominating.

Although broad conceptual frameworks, such as the Kapteyn Universe or Shapley's "big-universe," organized the data, astronomers and cosmologists focused on empirical and conceptual problems whose solutions were driven by model building. In their quest for answers, astronomers were strongly motivated to seek out and encourage the development of new technologies. In the

---

[26] See, for example, Dimitri Mihalas and Paul M. Routly, *Galactic Astronomy* (San Francisco: W. H. Freeman and Co., 1968), pp. 47–87.

period defined by this study, astronomy was radically influenced by the development of spectroscopy, photometry, and, of course, the newer class of reflecting telescopes, particularly at the Mount Wilson Observatory. These two essential features of the modern scientific enterprise worked in a symbiotic fashion to shape a cosmology during the 1920s that remains one of the dominant intellectual achievements of the twentieth century.

APPENDIX I
SEELIGER'S STAR-RATIO FUNCTION

Seeliger modified the fundamental stellar theorem with the empirical results obtained from his second stellar relationship in order to derive his star-ratio function:

$$A_m = ch_m^{(\lambda-3)/2}, 0 < \lambda < 1,$$

where

$$
\begin{aligned}
A_m &\equiv \text{star-ratio function with magnitude } m; \\
c &\equiv \text{constant of proportionality;} \\
h_m &\equiv \text{apparent brightness of a star of magnitude } m; \\
\lambda &\equiv \text{empirical quantity independent of } h_m.
\end{aligned}
$$

The fundamental stellar theorem defines the relationship between the star-ratio function and the apparent brightness of a star with magnitudes m and n, that is

$$\frac{A_m}{A_n} = \sqrt[3]{\frac{h_n}{h_m}}. \tag{1}$$

The difference between successively fainter magnitude classes is defined by Pogson's light-ratio ($= 2.512$), or

$$
\begin{aligned}
\log h_m - \log h_{m+1} &= \log(2.512) \\
&= 0.4.
\end{aligned}
$$

For half-magnitude differentials, we find

$$
\begin{aligned}
\log \frac{h_m}{h_{m+1/2}} &= \log h_m - \log h_{m+1/2} \\
&= \tfrac{1}{2}(\log h_m - \log h_{m+1}) \\
&= \tfrac{1}{2}(0.4) \\
&= 0.2
\end{aligned}
$$

Letting $\alpha = \frac{A_{m+1/2}}{A_m}$, and substituting into Equation 1, we find,

$$
\begin{aligned}
\log \alpha &= \log \frac{A_{m+1/2}}{A_m} \\
&= \log(\frac{h_m}{h_{m+1/2}})^{\frac{3}{2}} \\
&= \tfrac{3}{2} \log \frac{h_m}{h_{m+1/2}} \\
&= \tfrac{3}{2}(0.2) \\
&= 0.3.
\end{aligned}
$$

This last value is the theoretical constant of the star-ratio function under the conditions specified by the fundamental stellar theorem. On the basis of empirically derived star-ratios for each of the galactic zones, Seeliger's second stellar relationship indicated that this value was not confirmed. In tabular form, he found:

zone	$\log \alpha_o$	λ
I	0.237	+0.63
II and VIII	0.243	+0.57
III and VII	0.248	+0.52
IV and VI	0.260	+0.40
V	0.275	+0.25

where

$$0 < \lambda = \log \alpha - \log \alpha_o < 1,$$

or

$$\log \alpha_o = \log \alpha - \lambda.$$

In other words, the theoretical value of the star-ratio value ($=0.3$) should be modified by λ. This implies that the exponent must be changed from $\frac{3}{2}$ to $\frac{3-\lambda}{2}$. Consequently,

$$\log \alpha_o = \log(\frac{h_{m+1/2}}{h_m})^{(3-\lambda)/2} \tag{2}$$

where λ is defined in the table above. Because

$$\alpha_o = \frac{A_m}{A_{m+1/2}},$$

we find equating this with Equation (2),

$$\log \frac{A_m}{A_{m+1/2}} = \log(\frac{h_{m+1/2}}{h_m})^{(3-\lambda)/2}$$

$$= \log(\frac{h_m}{h_{m+1/2}})^{(\lambda-3)/2}.$$

Therefore,

$$A_m = c h_m^{(\lambda-3)/2}.$$

APPENDIX II
SEELIGER'S DENSITY THEOREM

Given his modified star-ratio function, Seeliger derived his density theorem first in 1898:

$$\text{If } A_m = ch_m^{(\lambda-3)/2}, \text{ then } D = \gamma r^{-\lambda}.$$

where:

A_m	$\equiv$	star-ratio function with magnitude m;
c	$\equiv$	constant of proportionality;
h_m	$\equiv$	apparent brightness of a star of magnitude m;
λ	$\equiv$	empirical quantity independent of h_m;
γ	$\equiv$	empirically derived numerical quantity;
D	$\equiv$	density function;
r	$\equiv$	radial distance from the Sun;
i	$\equiv$	absolute brightness;
H	$\equiv$	magnitude of the brightest star (apparent brightness at the upper limit);
$\Phi(i)$	$\equiv$	frequency of occurrence (i.e., percentage of stars) of the given brightness i ($0 \leq i \leq H$);
$\Phi(i)di$	$\equiv$	percentage of stars (i.e., frequency of occurrence) whose absolute brightness lies between i and $i+di$.

Assume the totality of all stars (i.e., across all brightnesses) is equal to 1, or

$$\int_0^H \Phi(i) \, di = 1. \tag{1}$$

For an arbitrary distance, r, from the Sun,

$$i = hr^2. \tag{2}$$

Let

r_c	$\equiv$	unit distance (i.e., $r = 1$) where the absolute brightness equals the apparent brightness (i.e., $i = h$);
dt	$\equiv$	a differential unit of volume (i.e., infinitely small) called a volume-element at the unit distance;

Ddt $\equiv$ density of stars (i.e., number of stars) contained in a volume-element;

$A(dt)$ $\equiv$ number of stars contained in a volume-element with a brightness between the limits i and $i+dt$.

Therefore,

$$A(dt) = Ddt\Phi(i)di. \tag{3}$$

From Equations 2 and 3, we get,

$$A(dt) = Ddt\Phi(hr^2)r^2dh. \tag{4}$$

Let

w $\equiv$ the apparent area of the stellar zone described by the unit distance.

At an arbitrary distance, w describes an area equal to wr^2. Therefore, the volume of the corresponding area of an infinitely thin spherical shell is

$$dt = wr^2dr. \tag{5}$$

Combining Equations 4 and 5, we get,

$$A(dt) = Dwr^4\Phi(hr^2)drdh, \tag{6}$$

which expresses analytically the number of stars, $A(dt)$, contained in an infinitely thin area (w), at a given distance (r), with apparent brightness between limits h and $h + dh$. Integrating Equation 6 yields the actual number of stars (A_m) between the specified limits of brightness and distance.

The limits of integration are determined from the following considerations. Because Equation 6 contains the differential of apparent brightness, dh, we must first determine the limits of apparent brightness. From Equation 2 we find that the brightest star, H, in this volume has an apparent brightness $h = H/r^2$, whereas the faintest star of magnitude m has an apparent brightness h_m. Therefore, first integrating with respect to brightness (magnitude) we get,

$$A_m(dt) = Dwr^4dr \int_{h_m}^{H/r^2} \Phi(hr^2)\,dh. \tag{7}$$

Integrating over distance in the direction of w, we eliminate dt, which is an infinitely thin slice of the cone emanating from the Sun. Let r be the distance to the nearest star. Seeliger assumed that because relatively few stars exist in the immediate neighborhood of the Sun, the approximation $r' = 0$ can be accepted for the lower limit. The upper limit is the distance required to include all the stars under consideration. Because H represents the brightest star, we find the upper limit from Equation 2,

$$H = h_m r^2,$$

or

$$r = \sqrt{H/h_m}.$$

Integrating Equation 7 with respect to distance, we get

$$A_m = w \int_0^{\sqrt{H/h_m}} Dr^4 dr \int_{h_m}^{H/r^2} \Phi(hr^2) dh. \tag{8}$$

As we have seen, the limits of apparent brightness are

$$h_m \leq h \leq H/r^2.$$

Multiplying by the factor r^2 we obtain the limits of absolute brightness,

$$h_m r^2 \leq hr^2 \leq H,$$

or

$$h_m r^2 \leq i \leq H.$$

Making the substitution $x = hr^2$, and transforming Equation 8 to represent the limits of absolute brightness, we get

$$A_m = w \int_0^{\sqrt{H/h_m}} Dr^2 dr \int_{h_m r^2}^{H} \Phi(x) dx. \tag{9}$$

Separating Equation 9, we obtain

$$A_m = w \int_0^{\sqrt{H/h_m}} Dr^2 u\, dr, \tag{10}$$

and

$$u = \int_{h_m r^2}^{H} \Phi(x) dx. \tag{11}$$

APPENDIX II

Utilizing these last two equations, Seeliger derived a series of necessary relations. From equation 11, we find,

$$du = \Phi(x)dx,$$

or

$$\frac{du}{dx} = \Phi(x).$$

Differentiating Equation 2 with respect to h and $r(x = i)$ separately, we get,

$$\frac{dx}{dh_m} = r^2$$

and

$$\frac{dx}{dr} = 2rh_m.$$

Utilizing the "chain rule of differentiation,"

$$\frac{du}{dh_m} = \frac{du}{dx}\frac{dx}{dh_m},$$

we find,

$$\frac{du}{dh_m} = \Phi(x)r^2. \tag{12}$$

Moreover, since

$$\frac{du}{dr} = \frac{du}{dx}\frac{dx}{dr},$$

we find similarly,

$$\frac{du}{dr} = \Phi(x)2rh_m. \tag{13}$$

Combining Equations 12 and 13, we get

$$\frac{du}{dh_m} = \frac{du}{dr}\frac{r}{2h_m}. \tag{14}$$

Finally, differentiating Equation 10 with respect to h_m, and substituting Equation 14, we obtain the relation

$$\frac{dA_m}{dh_m} = \frac{w}{2h_m} \int_0^{\sqrt{H/h_m}} Dr^3 \frac{du}{dr} dr. \tag{15}$$

Equation 15 can be reduced by integrating the right-hand portion by the "method of parts." Thus we find,

$$\frac{dA_m}{dh_m} = -\frac{w}{2h_m} \int_0^{\sqrt{H/h_m}} u(r^3 \frac{dD}{dr} + 3r^2 D)dr$$

Substituting Equation 11, we get,

$$= -\frac{3w}{2h_m} \int_0^{\sqrt{H/h_m}} Dr^2 u dr - \frac{w}{2h_m} \int_0^{\sqrt{H/h_m}} r^3 \frac{dD}{dr} dr \int_{h_m r^2}^{H} \Phi(x)dx.$$

Substituting Equation 10, we find,

$$= -\frac{3}{2h_m} A_m - \frac{w}{2h_m} \int_0^{\sqrt{H/h_m}} r^3 \frac{dD}{dr} dr \int_{h_m r^2}^{H} \Phi(x)dx. \tag{16}$$

The solution of this last equation depends entirely on the determination of three relationships: the star-ratio function (A_m), the density function (D), and the frequency-luminosity function (Φ). If any two of these relationships are known, then the third function can be obtained. Because Seeliger's purpose ultimately was to determine the density function, the Φ−function must be non-zero, otherwise the total illuminating power of the Universe would be zero. The star-ratio function (A_m) was given by the empirical relationship he had already derived from his second stellar law (see Appendix I).

The star-ratio function analytically states

$$A_m = c h_m^{(\lambda-3)/2}, 0 < \lambda < 1, \tag{17}$$

where λ and h_m are independent of one another.

Differentiating Equation 17 with respect to h_m, we obtain

$$\frac{dA_m}{dh_m} = c \frac{\lambda - 3}{2} \frac{h_m^{(\lambda-3)/2}}{h_m}$$

$$= \frac{\lambda - 3}{2} \frac{A_m}{h_m}$$

244

$$= -\frac{3}{2h_m}A_m + \frac{\lambda}{2h_m}A_m. \tag{18}$$

Equating 16 and 18, and substituting Equations 10 and 11, we get,

$$\int_0^{\sqrt{H/h_m}} (\lambda D r^2 + r^3 \frac{dD}{dr}) dr \int_{h_m r^2}^{H} \Phi(x) dx = 0.$$

Regardless of the actual value of the frequency-luminosity function (Φ), the latter equation will have a valid solution, if

$$\lambda D r^2 + r^3 \frac{dD}{dr} = 0.$$

Hence,

$$D = \gamma r^{-\lambda}.$$

BIBLIOGRAPHICAL NOTE

The largest collection of Kapteyn's correspondence is presently preserved in the George Ellery Hale Microfilm Collection, which contains several hundred letters between Kapteyn and Hale. Additional correspondence between Kapteyn and various astronomers, scientists, scholars, and others also exists in a variety of European and American libraries and archives, principally the Astronomical Laboratory at Groningen, the University of Groningen, and the Yerkes Observatory at Williams Bay, Wisconsin.

Perhaps equally important, the biography *J. C. Kapteyn: Zijn Leven en Werken* (Groningen: P. Noordhoff, 1928) by his daughter Henrietta Hertzsprung-Kapteyn reproduces many biographical and technical details from Kapteyn's correspondence with numerous scientists and scholars that have otherwise become unavailable as a result of the destruction of Kapteyn's personal archive during the bombing of Rotterdam in May 1940. Therefore, access to the Kapteyn biography becomes an archival treasure for future studies dealing with Kapteyn himself, as well as the history of both modern and Dutch astronomy, George Ellery Hale, Mount Wilson, and the rise of American astronomy. For an annotation of the Kapteyn biography, see my *The Life and Works of J. C. Kapteyn: An Annotated Translation* (Dordrecht: Kluwer Academic Publ., forthcoming).

Most of Seeliger's personal archive was also destroyed during the Second World War. Of the extant materials, Seeliger's most extensive correspondence was with Max Wolf, which is presently preserved at the Heidelberg University Library archives (sixty-seven items), and with Karl Schwarzschild, which is preserved in microfilm and available at the American Philosophical Society, the Center for the History of Physics at the American Institute of Physics, and the California Institute of Technology. Additional materials are located in the archives of the Munich Observatory and the Deutsches Museum (Munich). Other archival sources that were consulted in the preparation of this study are listed in the Acknowledgments.

Although many scholars have contributed broadly to the history of astronomy, only a small number have contributed directly to stellar and galactic studies, the subdiscipline that is the subject of the present study. I am excluding the many fine studies in twentieth-century astronomy that do not deal directly with galactic theory or stellar astronomy, such as those on radio and X-ray astronomy. Thus with only a few recent exceptions, scholarly literature in the history of astronomy has generally emphasized planetary astronomy to the neglect of those developments in stellar astronomy.

Within the broad field of stellar astronomy, statistical cosmology has been among those areas that have fallen victim to neglect. The primary literature upon which much of the present study is based is not re-produced here as it is provided within the footnotes. For a detailed discussion of nearly 500 primary research papers in this field, however, see my *Seeliger, Kapteyn and the Rise of Statistical Astronomy* (Ph.D. dissertation, Bloomington, Ind., 1976). Bibliographies of primary source scientific (astronomical) papers relating to every phase of the development of astronomy and astrophysics during the period of this study are found in *Astronomischer Jahrsbericht* (1889–1968) and *Astronomy and Astrophysics* (1969–present).

Among secondary studies in stellar astronomy and galactic theory, four publishing events have emerged over the course of the last several decades that have begun to reshape the discipline of the history of modern astronomy. With its first volume issued in 1970, the *Journal for the History of Astronomy* has begun to redress this imbalance. Although scholarly articles dealing with planetary astronomy have continued to appear, the *JHA* has also emphasized studies in galactic theory and stellar astronomy in general. Its editor, Professor Michael A. Hoskin, has almost single-handedly contributed to stellar studies himself.

Second, a variety of recent studies in stellar astronomy, particularly Professor Hoskin's *William Herschel and the Construction of the Heavens* (London: Oldbourne Book Co., 1963) and his *Stellar Astronomy: Historical Studies* (Chalfont St Giles: Science History Publ., 1982), along with Richard Berendzen, Richard Hart, and Daniel Seeley, *Man Discovers the Galaxies* (New York: Science History Publ., 1976) and Robert W. Smith's *The Expanding Universe: Astronomy's "Great Debate," 1900–1931* (Cambridge: Cambridge University Press, 1982), have reassessed much of the discipline by emphasizing the importance of unpublished sources.

Third, several synthetic studies of galactic theory and stellar astronomy have contributed to this new historiographic orientation, such as Dieter Herrmann's *The History of Astronomy from Herschel to Hertzsprung* (Cambridge: Cambridge University Press, 1984) and several volumes of

The General History of Astronomy under the general editorship of
Professor Hoskin published by Cambridge University Press. From the
latter a variety of relevant essays may be found in *Astrophysics and
Twentieth-Century Astronomy to 1950, Part A* (Cambridge: Cambridge
University Press, 1984), volume 4, edited by Owen Gingerich.

And lastly, a number of studies by historians of astronomy (and a few
historically oriented astronomers) have appeared in recent years in the
professional literature with increasing regularity; those relevant to and
used within the present study are listed below.

SECONDARY LITERATURE

Batten, A. H. *Resolute and Undertaking Characters: The Lives of Wilhelm
and Otto Struve* (Dordrecht: D. Reidel Publ., 1988).
Bennett, J. A. "On the Power of Penetrating into Space: The Telescopes
of William Herschel," *Journal for the History of Astronomy*, 7(2)
(1976), 75–108.
Berendzen, Richard, and Michael Hoskin. "Hubble's Announcement of
Cepheids in Spiral Nebulae, " *Astronomical Society of the Pacific,
Leaflet*, no. 504 (1971).
Berendzen, Richard, and Richard Hart. "Adriaan van Maanen's Influ-
ence on the Island Universe Theory," *Journal for the History of
Astronomy*, 4(1) (1973), 46–56, 4(2) (1973), 73–98.
Berendzen, Richard, Richard Hart, and Daniel Seeley. *Man Discovers
the Galaxies* (New York: Science History Publ., 1976).
Bertolli, B., Balbinot, Bergia, and Messnia, eds. *Modern Cosmology in
Retrospect* (Cambridge: Cambridge University Press, 1990).
Bok, Bart. "Harlow Shapley–Cosmographer and Humanitarian," *Sky and
Telescope*, 44 (1972), 354–7.
Bok, Bart. "Harlow Shapley and the Discovery of the Center of Our
Galaxy," in Jerzy Newman, ed., *The Heritage of Copernicus*
(Cambridge, Mass.: MIT Press, 1974), pp. 26–62.
Bok, Bart. "The Universe Today," in D. W. Corson, ed., *Man's Place in
the Universe: Changing Concepts* (Tuscon: University of Arizona
Press, 1977), pp. 95–139.
Brush, Stephen G. "Thermodynamics and History: Science and Culture
in the 19th Century," *Graduate Journal*, 7 (1967), 477–565, re-
printed as *The Temperature of History: Phases of Science and
Culture in the Nineteenth Century* (New York: B. Franklin, 1978).
Brush, Stephen G. "A Geologist among Astronomers: The Rise and Fall
of the Chamberlin–Moulton Cosmogony," *Journal for the History
of Astronomy*, 9(1) (1978), 1–41, 9(2) (1978), 77–104.

Crowe, Michael J. *The Extraterrestrial Life Debate, 1750–1900: The Idea of a Plurality of Worlds from Kant to Lowell* (New York: Cambridge University Press, 1986).

DeVorkin, David H. "The Origins of the Hertzsprung–Russell Diagram," in A. G. Davis and David H. DeVorkin, eds., *In Memory of Henry Norris Russell* (Dudley Observatory Report no. 13, 1977), pp. 61–77.

DeVorkin, David H. "A Sense of Community in astrophysics: Adopting a System of Spectral Classification," *Isis*, 72 (261) (1981), 29–49.

DeVorkin, David. "Stellar Evolution and the Origin of the Hertzsprung–Russell Diagram," in Owen Gingerich, ed., *Astrophysics and Twentieth-Century Astronomy to 1950. Part A* (Cambridge: Cambridge University Press, 1984), pp. 90–108.

Dick, Steven J., ed. "National Observatories: An Overview," *Journal for the History of Astronomy*, 22(1) (1991), 1–99.

Dieke, Sally H. "Karl Schwarzschild," in C. C. Gillispie, ed., *Dictionary of Scientific Biography* (New York: Charles Scribner's & Sons, 1975, 16 vols.), vol. 12, pp. 247–53.

Donovan, Arthur, Larry Laudan, and Rachael Laudan, eds. *Scrutinizing Science: Empirical Studies of Scientific Change* (Dordrecht: D. Reidel Publ., 1988).

Douglas, A. V. *The Life of Arthur Stanley Eddington* (London: Nelson, 1956).

Dreyer, J. L. E., and H. H. Turner, eds. *History of the Royal Astronomical Society, 1820–1920*, vol. 1 (Oxford: Blackwell Publ., 1987).

Dupree, A. Hunter, ed. "Smithsonian Astrophysical Observatory Centennial," *Journal for the History of Astronomy*, 21(1) (1990), 107–53.

Eisberg, Joann. "Eddington's Stellar Models and Early Twentieth-Century Astrophysics" (Ph.D. dissertation, Harvard University, 1991).

Elliott, Clark A. "Harvard College Observatory Sesquecentennial," *Journal for the History of Astronomy*, 21(1) (1990), 3–106.

Fleck, G. M. "Atomism in Late 19th Century Physical Chemistry," *Journal for the History of Ideas*, 24 (1963), 109–110.

Flexner, Abraham. *Henry S. Pritchett: A Biography* (New York: Columbia University Press, 1943).

Fox, R. "The Rise and Fall of Laplacian Physics," *Historical Studies in the Physical Sciences*, 4 (1974), 89–136.

Frank, P. "The Mechanical Versus the Mathematical Conception of Nature," *Philosophy of Science*, 4 (1937), 41–74.

Gingerich, Owen. "Harlow Shapley," in C. C. Gillispie, ed., *Dictionary of Scientific Biography* (New York: Charles Scribner's & Sons, 1975, 16 vols.), vol. 12, pp. 345–52.

Gingerich, Owen, ed. *Astrophysics and Twentieth-Century Astronomy to 1950, Part A* (Cambridge: Cambridge University Press, 1984).

Gingerich, Owen, ed. "Astronomical Institutions," in Owen Gingerich, ed., *Astrophysics and Twentieth-Century Astronomy to 1950: Part A* (Cambridge: Cambridge University Press, 1984), pp. 111–33.

Gingerich, Owen. "How Shapley Came to Harvard or, Snatching the Prize from the Jaws of Debate," *Journal for the History of Astronomy*, 19(3) (1988), 201–7.

Heilbron, John. *The Dilemmas of an Upright Man: Max Planck as Spokesman for German Science* (Berkeley: University of California Press, 1986).

Herrmann, Dieter. "B. A. Gould and His *Astronomical Journal*," *Journal for the History of Astronomy*, 2(2) (1971), 98–108.

Herrmann, Dieter. "N. R. Pogson and the Definition of the Astrophotometric Scale," *Journal of the British Astronomical Association*, 87 (1977), 146–9.

Herrmann, Dieter. *The History of Astronomy from Herschel to Hertzsprung* (Cambridge: Cambridge University Press, 1984).

Hetherington, Norris. "Adriaan van Maanen and Internal Motions in Spiral Nebulae: A Historical Review," *Quarterly Journal of the Royal Astronomical Society*, 13 (1972), 25–39.

Hetherington, Norris. "Adriaan van Maanen on the Significance of Internal Motions in Spiral Nebulae," *Journal for the History of Astronomy*, 5(1) (1974), 57–8.

Hetherington, Norris. "Edwin Hubble on Adriaan van Maanen's Internal Motions in Spiral Nebulae," *Isis*, 65 (1974), 390–3.

Hetherington, Norris. "The Simultaneous 'Discovery' of Internal Motions in Spiral Nebulae," *Journal for the History of Astronomy*, 6(2) (1975), 115–25.

Hetherington, Norris. *Science and Objectivity* (Ames: Iowa State University Press, 1988).

Hetherington, Norris. "Hubble's Cosmology," *American Scientist*, 78 (1990), 142–51.

Hiebert, Edwin H. "The Energetics Controversy and the New Thermodynamics," in D. H. D. Rooler, ed., *Perspectives in the History of Science and Technology* (Norman: University of Oklahoma Press, 1971), pp. 67–86.

Hoskin, Michael A. *William Herschel and the Construction of the Heavens* (London: Oldbourne Book Co., 1963).

Hoskin, Michael A. "The 'Great Debate': What Really Happened," *Journal for the History of Astronomy*, 7(3) (1976), 169–82.

Hoskin, Michael A. "Newton, Providence and the Universe of Stars," *Journal for the History of Astronomy*, 8(2) (1978), 77–101.

Hoskin, Michael A. "William Herschel's Early Investigations of Nebulae: A Reassessment," *Journal for the History of Astronomy*, 10 (3) (1979), 165–76.

Hoskin, Michael A. *Stellar Astronomy: Historical Studies* (Bucks, England: Science History Publ., 1982).

Hoskin, Michael A. "Astronomical Correspondence of William Rowan Hamilton," *Journal for the History of Astronomy*, 15 (1984), 69–73.

Hoskin, Michael A. "John Herschel's Cosmology," *Journal for the History of Astronomy*, 18(1) (1987), 1–34.

Hufbauer, Karl. *Exploring the Sun: Solar Science since Galileo* (Baltimore: The Johns Hopkins University Press, 1991).

Jaki, Stanley L. *The Paradox of Olbers' Paradox: A Case History of Scientific Thought* (New York: Herder and Herder, 1969).

Jaki, Stanley L. *The Milky Way: An Elusive Road for Science* (New York: Science History Publ., 1972).

Jaki, Stanley L. "Das Gravitations-Paradoxon des unendlichen Universums," *Sudhoffs Archiv für Geschichte der Medizin und der Naturwissenschaften*, 63 (1979), 105–22.

Jaki, Stanley L. *Cosmos in Transition: Studies in the History of Cosmology* (Tucson: Pachart Publ., 1990).

Jungnickel, Christa, and Russell McCormmach. *Intellectual Mastery of Nature: Theoretical Physics from Ohm to Einstein*, 2 vols. (Chicago: University of Chicago Press, 1986).

Kevles, Daniel J. "'Into Hostile Political Camps': The Reorganization of International Science in World War I," *Isis*, 62 (1971), 47–60.

Kline, Morris. "Mechanical Explanation at the End of the 19th Century," *Centaurus*, 17 (1972), 58.

Krisciunas, Kevin. *Astronomical Centers of the World* (Cambridge: Cambridge University Press, 1988).

Lankford, John. "The Impact of Photography on Astronomy," in Owen Gingerich, ed., *Astrophysics and Twentieth-Century Astronomy to 1950, Part A* (Cambridge: Cambridge University Press, 1984), vol. 4, pp. 16–39.

Laudan, Larry. *Progress and Its Problems: Toward a Theory of Scientific Growth* (Berkeley: University of California Press, 1977).

Laudan, Larry. *Science and Values* (Berkeley: University of California Press, 1984).

Merz, John T. *A History of European Thought in the Nineteenth Century* (New York: Dover Publ., 1965).

Nielsen, A. V. "Contributions to the History of the Hertzsprung–Russell Diagram," *Centaurus* ix (1964), 219–53.

North, John D. *The Measure of the Universe: A History of Modern Cosmology* (Oxford: Clarenden Press, 1965).

Osterbrock, Donald E. "Failure and Success: Two Early Experiments with Concave Gratings in Stellar Spectroscopy," *Journal for the History of Astronomy*, 17(2) (1986), 119–29.

Osterbrock, Donald E. *James E. Keeler: Pioneer American Astrophysicist and the Early Development of American Astrophysics* (Cambridge: Cambridge University Press, 1987).

Osterbrock, Donald E., John R. Gustafson, and W. J. Shiloh Unruh, *Eye on the Sky: Lick Observatory's First Century* (Berkeley: University of California Press, 1988).

Pannekoek, Anton. *A History of Astronomy* (New York: Barnes and Noble, 1961).

Paul, E. Robert. "The Nature of the Nebulae," *Journal for the History of Astronomy*, 9(3) (1978), 222–4.

Paul, E. Robert. "The Death of a Research Programme: Kapteyn and the Dutch Astronomical Community," *Journal for the History of Astronomy*, 12(2) (1981), 77–94.

Paul, E. Robert. "Festschrift for Oort," *Journal for the History of Astronomy*, 13(2) (1982), 141–2.

Paul, E. Robert. "Kapteyn and Statistical Astronomy," in Hugo van Woerden, Ronald J. Allen, and W. Butler Burton (eds.), *The Milky Way Galaxy* (Dordrecht: D. Reidel Publ., 1984), pp. 25–42.

Paul, E. Robert. "J. C. Kapteyn and the Early Twentieth-Century Universe," *Journal for the History of Astronomy*, 17(3) (1986), 155–82.

Plotkin, Howard. "Edward C. Pickering, the Henry Draper Memorial, and the Beginnings of Astrophysics in America," *Annals of Science*, 35 (1978), 365–77.

Rohlfs, Kristen. "Galactic Astronomy in Continental Europe in the 19th Century in the time succeeding Herschel," *Vistas in Astronomy*, 32 (1989), 215–23.

Sandage, Allan. "Edwin Hubble, 1889–1953," *Journal of the Royal Astronomical Society of Canada*, 83 (1989), 351–62.

Schmeidler, Felix. *Die Geschichte der Astronomischen Gesellschaft* (Hamburg: Klare Verlag, 1988).

Seeley, Daniel, and Richard Berendzen. "The Development of Research in Interstellar Absorption, c. 1900–1930," *Journal for the History of Astronomy*, 3(1) (1972), 52–64, 3(2) (1972), 75–86.

Seeley, Daniel, and Richard Berendzen. "Astronomy's Great Debate," *Mercury*, 7 (1976), 67–71, 88.

Shapley, Harlow. *Through Rugged Ways to the Stars* (New York: Charles Scribner's & Sons, 1969).

Sheynin, O. B. "On the History of the Statistical Method in Astronomy," *Archive for History of Exact Sciences*, 29 (1984), 151–99.

Silliman, R. H. "Fresnel and the Emergence of Physics as a Discipline," *Historical Studies in the Physical Sciences*, 4 (1974), 137–62.

Smith, Robert W. *The Expanding Universe: Astronomy's "Great Debate," 1900–1931* (Cambridge: Cambridge University Press, 1982).

Smith, Robert W. "Edwin P. Hubble and the Transformation of Cosmology," *Physics Today*, 43(4) (1990), 52–8.

Struve, Otto. "A Historic Debate about the Universe," *Sky and Telescope*, 19 (1960), 398–401.

Struve, Otto. "M. A. Kovalsky and His Work on Stellar Statistics," *Sky and Telescope*, 23 (1962), 251–2.

Struve, Otto, and V. Zebergs. *Astronomy of the Twentieth Century* (New York: Macmillan & Co., 1962).

Struve, Otto. *The Universe* (Cambridge, Mass.: MIT Press, 1964).

Szanser, A. J. "Marian Kowalski (1821–1884): A Little Known Pioneer in Stellar Statistics," *Royal Astronomical Society, Quarterly Journal*, 11 (1970), 343–7.

Szanser, A. J. M. "F. G. W. Struve (1793–1864). Astronomer at the Pulkovo Observatory," *Annals of Science*, 28 (1972), 338–343.

Thoren, Victor E., Charles Gow, and Kent Honeycutt. "An Early View of Galactic Rotation," *Centaurus*, 18 (1974), 301–14.

Tucker, Wallace, and Karen Tucker. *The Cosmic Inquirers: Modern Telescopes and the Makers* (Cambridge, Mass.: Harvard University Press, 1986).

Turchetta, M., and G. Gavazzi. "Nineteenth-Century Italian Contributions to Galactic Theory," *Journal for the History of Astronomy*, 18(3) (1987), 196–208.

Warner, Brian. *Astronomers at the Royal Observatory, Cape of Good Hope: A History with Emphasis on the 19th Century* (Cape Town: Balkema for the University of Cape Town, 1979).

Whitney, Charles. *The Discovery of Our Galaxy* (New York: Alfred Knopf, 1972).

Whitrow, G. J. "From the Problem of Fall to the Problem of Collapse: Three Hundred Years of Gravitational Theory," *Philosophical*

Journal: Transactions of the Royal Philosophical Society of Glascow, 14 (1977), 67–84.

van de Kamp, Peter. "The Galactocentric Revolution, a Reminiscent Narrative," *Astronomical Society of the Pacific, Publications*, 77 (1965), 325–35.

van Woerden, H., W. N. Brouw, and H. C. van de Hulst, eds. *Oort and the Universe* (Dordrecht: D. Reidel Publ., 1980).

Wright, Helen. *Explorer of the Universe: A Biography of George Ellery Hale* (New York: Dutton & Co., 1966).

Wright, Helen. "Adams, Walter Sydney," in C. C. Gillispie, ed., *Dictionary of Scientific Biography* (New York: Charles Scribner's & Sons, 1975, 16 vols.), vol. 1, pp. 54–8.

Wright, Helen. "Hale, George Ellery," in C. C. Gillispie, ed., *Dictionary of Scientific Biography* (New York: Charles Scribner's & Sons, 1975, 16 vols.), vol. 4, pp. 26–34.

Wright, Helen, J. Wurnaw, and C. Weiner. *The Legacy of George Ellery Hale* (Cambridge, Mass.: MIT Press, 1972).

Wright, Helen. *James Lick's Monument* (New York: Cambridge University Press, 1987).

INDEX

Abbot, Charles G., 167, 218; *see also* National Academy of Sciences

Adams, Walter S., 103; and George Ellery Hale, 103, 170, 179; giant stars, 219; and J. C. Kapteyn, 152, 170, 204n.41; Mt. Wilson Observatory, 103, 113, 143; *Plan of Selected Areas*, 166; and Hugo von Seeliger, 179; spectroscopic parallaxes, 113; stellar absorption, 103–6, 109, 111, 143

Airy, George Biddell, 25, 46

American Astronomical Society, 182, 221

Amsterdam Academy of Sciences, 85, 166

Anding, F., 177

Andromeda Nebulae (M33), *see* stellar objects

anti-German, 171, 172, 175; *see also* Manifesto of 93

Argelander, F. W. A., 45, 46, 65, 69; *Bonner Durchmusterung*, 32, 33, 63, 81; naked-eye star atlas (1843), 22, 31–2; see also *Bonner Durchmusterung des Nordlichen Himmels*

Astronomer Royal of England, 20, 132

Astronomical Journal, the, 183

Astronomical Laboratory at Groningen, *see* Kapteyn Astronomical Laboratory

Astronomical Society of the Pacific, 163, 182

Astronomische Gesellschaft, *see* Deutsche Astronomische Gesellschaft

Astronomische Nachrichten, 183

Astrophysical Journal, the, 183

Astrophysical Observatory at Potsdam, 168, 169, 175; *see also* Schwarzschild, Karl

atomic theory, 138

Auwers, G. F., 47, 84, 86

Baade, Walter, 202–3, 211

Babcock, Harold D., 103

Bailey, Solon I., 194

Baillaud, Jules, 161

Barnard, E. E., 178

Bauschinger, J., 177

Bentley, Richard, 34

Bessel, F. W., 31; 61 Cygni, 21, 45, 46; double star astronomy 41; *see also* Bessel–Weisse zone catalogue, Bradley, James

Bessel–Weisse zone catalogue, 22, 29

big galaxy, 51, 214–20, 235; *see also* Shapley, Harlow; twentieth-century cosmology

Bohlin, Karl, 197n.25

Bok, Bart, 142, 162

Bolzano, Bernhard, 67

Bonner Durchmusterung des Nordlichen Himmels (Argelander), 5, 31–42, 121; and J. C. Kapteyn, 47; and Edward Pickering 43; and Hugo von Seeliger, 58–64, 71–4; *see also* Argelander, F. W. A.

Bottlinger, K. F. E., 177

Bradley, James, on star catalogue, 21, 47, 86, 131

British Association for the Advancement of Science (BAAS), 89, 126, 175; *see also* star-streaming

Bruce Medal (Astronomical Society of the Pacific), *see* J. C. Kapteyn

Bruhns, E. H., and Hugo von Seeliger, 56, 65

Campbell, W. W., 3, 161–2, 182; and Heber D. Curtis, 202, 219; radial velocities, 89

Cannon, Annie Jump, and spectral classification, 108–9

Cantor, Georg, and set theory, 67–8

Cape of Good Hope, and BAAS, 89, 166; and David Gill, 81, 152, 166; and John Herschel, 29, 42

Cape Photographic Durchmusterung (Gill), 83, 162, 163

Carnegie Foundation for the Advancement of Teaching, 178

island universe theory, 49, 207, 212; confirmation of, 222; Kapteyn's support of, 202–3, 215; Shapley's rejection of, 197, 199–202; Shapley's support of, 195; Seeliger's support of, 215; *see also* Great Debate; twentieth-century cosmology

Jacobi, Carl Gustav Jacob, 56
Jeans, James H.: Kapteyn Universe, 151, 151n.26, 157–8, 211–12; on Maxwell's law, 127; *see also* star-streaming

Kant, Immanuel, on cosmology, 1, 14, 20
Kapteyn Astronomical Laboratory, 83, 161, 163; *see also Publications of the Astronomical Laboratory of Groningen*
Kapteyn, Jacobus Cornelius, 81; Bruce Medal (Astronomical Society of the Pacific), 163, 182; density function, 99, 131–2, 150–8; density galactic relationship, 110–11; distribution of stars, 80, 94–9, 112–26; Gaussian stellar parallaxes, 111–12; and David Gill, 81, 83; and Hugo Gyldén, 40; international science, 162–75; internationalizing of statistical cosmology, 181–5; interstellar absorption, 75, 101–6; Kapteyn Universe, 91n.20, 141–5, 150–9, 180, 182, 222–7; luminosity function, 100–1, 107–8; luminosity spectral relationship, 108–10; mean parallaxes, 47, 84, 94–9, 145; Mt. Wilson Observatory, 136, 204; Mt. Wilson research associate, 89–90, 101, 163, 167, 171, 184; old boys, 202–3, 205n.45, 206; *Order Pour le Mérite*, 174–5; proper motion and spectra 48–9; and Hugo von Seeliger, 176–7; Shapley's big galaxy, 189–91, 202–12; statistical cosmology, 2–4, 6, 7, 9, 13, 49–51, 55, 78, 160–2, 180, 221; stellar assumptions, 101–12; stellar statistics, 56, 58, 60, 85, 136–40, 176; sun-centered stellar system, 106–7; University of Groningen, 152; *see also* star-streaming
Kapteyn Universe, *see* Kapteyn, J. C.
Kapteyn, Willem, 85
Kelvin, Lord, 134
Kepler, Johannes, 130, 141
Kienle, Hans, 177; local system, 159, 216–18, 222; at Potsdam, 177–8; and

Hugo von Seeliger, 181, 215–16; and Harlow Shapley, 78, 158, 179–80, 183, 213–14, 217
kinetic theory of gases, 127, 138
Kline, Felix, 67
Kobold, Hermann: random stellar motions, 86; on Seeliger, 74, 79
Kohlschütter, Arnold: interstellar light absorption, 103–4; Mt. Wilson Observatory, 169–70; big galaxy, 211
Kopff, A., 207
Kovalsky, Marian A., and preferential stellar motion, 86
Kronecker, Leopold, 67
Küstner, K. F., 166

Lagrange, Joseph-Louis, 56
Lambert, Johann, on cosmology, 1, 14, 20
Laplace, Pierre Simon de, 56; universal gravitation, 69–70
Leavitt, Henrietta Swan, and period–luminosity relationship, 193–4, 196
Lindblad, Bertil: and C. V. L. Charlier, 122; differential galactic rotation, 131, 182, 189, 231; statistical cosmology, 211, 224–7, 235; twentieth-century cosmology, 8; *see also* Oort, Jan
Lindblad–Oort theory, *see* differential galactic rotation
Littrow, Karl L. von: and Benjamin A. Gould, 41; and Giovanni Schiaparelli, 44; and Hugo von Seeliger, 56, 59–62, 65–6; statistical astronomy, 36–8, 41, 50
local system of stars, 216–17, 222, 224, 230, 231
Lowell, Percival, 179, 219
Ludendorff, Hans, 170
luminosity function, 6, 80, 101; and C. V. L. Charlier, 125–6; and Arthur S. Eddington, 132; and Hugo Gyldén, 38–40, 66; and J. C. Kapteyn, 84, 95–9, 108–10, 111, 114, 132, 154–5, 205–6; luminosity-curve, 99, 143–4, 205–6; and Hugo von Seeliger, 76–7, 146–8
luminosity law, *see* luminosity function
Lunds Astronomiska Observatorium (Charlier), 184
Lusitania, 174
Luyten, Willem J., 207; and Harlow Shapley, 219